采后果实
品质劣变机理及调控研究

孙华军　李　跃　钱丽丽　著

中国农业科学技术出版社

图书在版编目(CIP)数据

采后果实品质劣变机理及调控研究 / 孙华军，李跃，钱丽丽著. --北京：中国农业科学技术出版社，2024.9.
ISBN 978-7-5116-7044-1

Ⅰ.S311

中国国家版本馆 CIP 数据核字第 2024WE1407 号

责任编辑	周丽丽
责任校对	李向荣
责任印制	姜义伟　王思文

出 版 者	中国农业科学技术出版社
	北京市中关村南大街 12 号　　邮编：100081
电　　话	(010) 82106638（编辑室） 　(010) 82106624（发行部）
	(010) 82109709（读者服务部）
网　　址	https://castp.caas.cn
经 销 者	各地新华书店
印 刷 者	北京虎彩文化传播有限公司
开　　本	170 mm×240 mm　1/16
印　　张	11.5
字　　数	200 千字
版　　次	2024 年 9 月第 1 版　2024 年 9 月第 1 次印刷
定　　价	68.00 元

◆◆◆ 版权所有·翻印必究 ◆◆◆

资助项目

黑龙江省自然科学基金联合引导项目"RiMADS调控细胞壁降解基因介导树莓果实软化的分子机制研究"（项目编号：LH2021C067）

黑龙江八一农垦大学引进人才科研启动基金项目"采后树莓果实软化的分子调控机理研究"（项目编号：XYB202102）

黑龙江八一农垦大学引进人才科研启动基金项目"基于RNA-seq发掘采后树莓花青素含量变化关键基因 研究"（项目编号：XYB202211）

前　　言

果蔬是人们日常生活中必不可少的食物，能为人类生命活动提供丰富的营养、能量和生物活性物质，包括维生素、氨基酸、蛋白质、不饱和脂肪酸、膳食纤维、胡萝卜素、矿物质、多酚、类黄酮等，可提高人体抗氧化活性和抗病能力，对人体健康十分重要。随着社会经济的快速发展，人们的健康意识日益增强，对果蔬的新鲜度要求也越来越严格，由于果蔬新鲜度与品质有关，且品质与营养特性呈正相关，因此果蔬品质的好坏直接影响到它们的市场竞争力。果蔬品质包括色泽、香气、口感、质地等感官品质，也包括可溶性糖、酸、维生素 C、纤维素等营养品质。

果蔬采后虽然失去了母体和土壤对其营养和水分的供给，但绝大多数的果蔬采后仍然是一个有生命的有机体，有较强的呼吸作用和旺盛的代谢速率，使得机体营养物质被大量消耗，品质下降，甚至腐烂变质，失去商品价值，影响经济效益。

果蔬采后品质下降过程中伴随着多种生理生化代谢活动变化，如能量代谢、糖代谢、活性氧代谢、膜脂代谢、细胞壁代谢等；还包括一系列分子水平的变化，如关键基因和转录因子表达水平的变化等。本书基于著者自 2019 年以来在果实采后品质变化的分子机理及调控研究方面积累的成果，着重介绍了采后果实品质劣变的生理生化和分子水平变化，以及延缓果实品质劣变的物理、化学和生物等方法，旨在为果蔬采后品质变化的机理及调控研究提供理论和实践指导。

"本书著者：孙华军（黑龙江八一农垦大学食品学院讲师，博士，硕士生导师）；李跃（黑龙江八一农垦大学食品学院讲师，博士）；钱丽丽（黑龙江八一农垦大学食品学院教授，博士，博士生导师）。全书共六章，其中第二、第四章和第六章及前言由孙华军执笔（约 10.2 万字），第一章第一、二、四节，第三章及第五章由李跃执笔（约 8.8 万字），第一章第三节由钱丽丽执笔。"

由于撰写时间仓促，书中难免有不足之处，恳请专家、同仁和广大读者批评指正。

著者

2024 年 8 月

目　　录

第一篇　采后果实感官和营养品质劣变机理研究

第一章　采后果实感官品质变化 ………………………………………… 3
　　第一节　色泽 ……………………………………………………………… 3
　　第二节　风味 ……………………………………………………………… 9
　　第三节　硬度 …………………………………………………………… 15
　　第四节　营养 …………………………………………………………… 17
第二章　采后果实生理生化变化 ………………………………………… 19
　　第一节　呼吸代谢 ……………………………………………………… 19
　　第二节　活性氧代谢 …………………………………………………… 20
　　第三节　能量代谢 ……………………………………………………… 22
　　第四节　膜脂代谢 ……………………………………………………… 22
　　第五节　细胞壁代谢 …………………………………………………… 42
　　第六节　碳水化合物代谢 ……………………………………………… 44
　　第七节　激素 …………………………………………………………… 45
第三章　采后果实分子水平变化 ………………………………………… 49
　　第一节　AP2/ERF 转录因子对采后果实品质的调控 ………………… 50
　　第二节　EIN3/EIL 转录因子对采后果实品质的调控 ………………… 50
　　第三节　MADS-box 转录因子对采后果实品质的调控 ……………… 51
　　第四节　MYB 转录因子对采后果实品质的调控 ……………………… 53
　　第五节　NAC 转录因子对采后果实品质的调控 ……………………… 54
　　第六节　其他转录因子对采后果实品质的调控 ……………………… 55

第二篇 采后果实的调控研究

- 第四章 物理保鲜处理对采后果实品质的影响 …………………………… 59
  - 第一节 低温处理对采后果实品质的影响 ……………………………… 59
  - 第二节 气调保鲜对采后果实品质的影响 ……………………………… 60
- 第五章 化学保鲜处理对采后果实品质的影响 …………………………… 62
  - 第一节 正丁醇对采后果实品质的影响 ………………………………… 62
  - 第二节 1-MCP 处理对采后果实品质的影响 ………………………… 63
  - 第三节 甜菜碱处理对采后果实品质的影响 …………………………… 64
  - 第四节 没食子酸丙酯对采后果实品质的影响 ………………………… 67
  - 第五节 氯化钙处理对采后果实品质的影响 …………………………… 69
  - 第六节 NO 处理对采后果实品质的影响 ……………………………… 72
- 第六章 生物保鲜处理对采后果实品质的影响 …………………………… 75
  - 第一节 褪黑素处理对采后果实保鲜的影响 …………………………… 75
  - 第二节 多胺处理对采后果实品质的影响 ……………………………… 90
  - 第三节 γ-氨基丁酸处理对采后果实品质的影响 ……………………… 93
  - 第四节 L-谷氨酸处理对采后果实品质的影响 ………………………… 95
  - 第五节 赤霉素处理对采后果实品质的影响 …………………………… 97
  - 第六节 糖处理对采后果实品质的影响 ………………………………… 98
  - 第七节 水杨酸处理对采后果实品质的影响 …………………………… 99
  - 第八节 抗坏血酸处理对采后果实品质的影响 ………………………… 102
  - 第九节 涂膜处理对采后果实品质的影响 ……………………………… 103
  - 第十节 其他处理对采后果实品质的影响 ……………………………… 104
- 参考文献 ………………………………………………………………………… 107

第一篇

采后果实感官和营养品质劣变机理研究

第一章　采后果实感官品质变化

第一节　色泽

果实感官品质包括色、香、味、质构等方面，一般通过人的各种感官判断。其中视觉方面的品质包括色泽、新鲜度、腐烂程度等，在消费者的初级选择中起着关键作用。成熟果实呈现的各种色泽主要三要素分为光源（可见光）、物体（物体微观结构与色素组成）和人眼（或者其他感光设备），而每一种色泽是由色相、明度及饱和度表示的（Wang et al., 2021），包括 L^*（亮暗，+表示偏白，-表示偏暗）、a^*（红绿，+表示偏红，-表示偏绿）、b^*（黄蓝，+表示偏黄，-表示偏蓝）、ΔE（总色差值）、$h°$（色相）等指标（谢章荟等，2024）。色素是来源于食材中的天然色素物质，具有抗氧化、增强免疫系统、抗菌、调节身体内环境等生理功能和可持续性、高生物降解性的特点（Fried et al., 2022）。天然色素在果蔬中呈现各种色泽或无色，环境变化产生的生化反应可引起色泽的变化。由于来源不同，同一种天然色素可能会有不同的颜色和强度。果蔬中的天然色素根据其结构分为四大类：异戊二烯衍生物、卟啉衍生物、多酚类、酮类及醌类衍生物（谢章荟等，2024）。

一、异戊二烯衍生物

异戊二烯衍生物类色素的分子间含有大量 C=C 的共轭体系，植物可通过类异戊二烯的次生代谢途径生成以异戊二烯为单位聚合而成的共轭双键长链色素，果蔬中以类胡萝卜素为代表（吕莹等，2023）。不考虑顺式和反式异构体的情况下，在食物中已经发现约 100 种类胡萝卜素（Ribeiro et al., 2021）。类胡萝卜素属于脂溶性的色素，在 400~500 nm 波长范围内吸光性能强，能呈现出红、橙、黄色。根据分子结构中是否含有氧原子，类胡萝卜素可分为：不含氧元素即只含碳氢元素的胡萝卜素，如番茄红素、α-胡萝卜素、β-胡萝卜素；

含氧元素的类胡萝卜素即其氧化衍生物，如玉米黄质、叶黄素、虾青素等。类胡萝卜素有助于许多水果和蔬菜的转黄，因此类胡萝卜素主要存在于深绿色或红黄色的果蔬中，如胡萝卜、甘薯、芒果、哈密瓜等，其特殊分子结构赋予了其减少氧化应激、保护视力、抑制癌细胞生长与增殖、保护心脏等多种重要功能（修伟业等，2023；高婧宇等，2024）。相关研究表明，光、热、氧、pH值等条件会使类胡萝卜素处于不稳定状态（Amorim et al., 2022），易氧化降解或异构化，在pH值和水分含量较低时受影响程度更高，从而造成采后果蔬贮藏过程中褪色和风味劣变（谢章荟等，2024）。

二、卟啉衍生物

卟啉是一类由四个吡咯类亚基的α-碳原子通过次甲基桥（=CH⁻）互联而形成的大分子杂环化合物，叶绿素是果蔬中最具代表性的卟啉衍生物类色素。叶绿素也是地球上含量最为丰富和重要的色素，可减少和清除自由基，防止脂质过氧化，安全性极高，因此广泛用于医疗、食品、化妆品等行业（郭艳华等，2015；孙婷等，2019；陈蓉等，2011）。高等植物中的叶绿素分为蓝绿色的叶绿素 a 和黄绿色的叶绿素 b，比例为 3∶1（温广宇等，2003）。叶绿素吸收大部分的红光和紫光但反射绿光因而呈现出绿色，在光合作用的光吸收中起核心作用。市面上常见的绿色果蔬（如猕猴桃、苦瓜、豌豆、芹菜等）中均含有丰富的叶绿素，绿色程度即叶绿素含量的高低往往能客观反映果蔬营养品质的优劣，叶绿素的降低是绿色植物对外界不利环境最敏感的反应。与类胡萝卜素类似，在果蔬加工过程中由于叶绿素的性质不稳定，热、光、酸（3.5~5.0）、氧、酶等环境因子也易使叶绿素发生降解反应生成一系列分解代谢物（胡宇微等，2023）。

果实在成熟过程中叶绿素逐渐消失，类胡萝卜素逐渐积累和显现出来。果实成熟时类胡萝卜素出现有两种情况：苹果、梨、香蕉等果实随着叶绿素降解，原有的类胡萝卜素不断显现出来；而番茄、柑橘等果实在成熟后期均有新合成的类胡萝卜素（李明启，1989；张海新等，2010）。采后绿色蔬菜叶绿素降解，呈现类胡萝卜素颜色，因此表现为绿色衰退，黄色增强。

三、多酚类

多酚类物质是果实生长代谢过程中的一种次生代谢产物，是所有含酚官能团物质的总称，因具有多个酚基团而得名，基本碳架结构组成是2-苯基苯并吡喃和多羟基（袁莹等，2018）。食品中多酚类化合物又称为人类"第七大营养素"（凌关庭，2000），具有抗氧化、降血糖、增强免疫功能、抑制细菌与癌细胞生长等作用（唐瑶等，2016）。多酚类物质可分为黄酮类、花青素类和单宁三大类（左玉，2013）。黄酮类物质的颜色范围为浅黄色至无色，这一类物质包括槲皮素、异黄酮类等。单宁也称鞣酸，具有涩味，能与金属离子或氧气反应而产生黑色。花青素是一类水溶性植物色素，包括花青素类色素、花黄素类色素和儿茶素类色素三种类型，广泛存在于多种深红色、紫色和蓝色的果蔬中，如浆果、红色水果及蔬菜等（Braga et al.，2018）。糖基化和酰基化产生了花青素的多样性（Rodriguez-Amaya，2019）。花青素易降解，pH值、氧气、温度和光照等环境因素都会影响它们的稳定性，导致褪色，形成无色或棕色化合物（Braga et al.，2018）。因此，以多酚类化合物为主要呈色色素的果蔬在加工过程中常发生酚类含量降低和色泽褐变的问题，成为在果蔬加工工业中应用的瓶颈（Zhang et al.，2019）。花色苷是花青素与糖分子通过糖苷键偶联而成的化合物，花色苷主要从植物材料中提取，但是提取的花色苷具有纯度低和性质不稳定等缺点，因此利用微生物代谢工程技术替代传统方法提取花色苷具有重要意义，著者曾对工程菌合成花色苷的研究进展，对酶的选择、运输调节、尿嘧啶核苷-5'-二磷酸葡萄糖供应量的调控以及工艺优化等方面进行了综述（李跃等，2020），主要内容如下。

(一) 花色苷生产时酶的筛选

由于不同植物中同源基因编码的酶在催化反应时表现出不同的动力学和热力学活性，当这些酶在外源微生物中表达时，会产生不同的代谢行为和目标化合物生产水平。因此，从不同物种中筛选酶是提高花色苷和黄酮类化合物产量的关键。Yan et al.（2008）对来自非洲菊、矮牵牛、金鱼草和苹果的异源ANS活性进行体外测试，并比较其体内催化效率，发现使用矮牵牛ANS在E. coli中制备花色苷时产量最高。同样，在E. coli中生产柚皮素时，通过不同同源物上游酶组合可以提高E. coli中柚皮素的产量，如将不同来源的4-香豆酰辅酶A的上游酶4-香豆酸辅酶A连接酶、柚皮苷查尔酮的上游酶CHS以及柚皮素的上游酶CHI组合，柚皮素产量发生显著变化（Jones et al.，2016）。采用该方法在E. coli中生产白藜芦醇时，白藜芦醇产量提高了2 300 mg/L（Lim et al.，2011）。

即使最适酶系统中酶的来源和组合均为最佳，在微生物生产花色苷时也不一定能够高产，这是因为这些植物来源的酶在原核细胞中属于异源表达。通常，这些酶的编码基因在表达之前需要经过一定的修饰。例如，P450还原酶的原核表达可在其 C 端融合来自长春花的 F3',5'H 片段，删除其 N 端膜锚定位点处的 4 个密码子，修改起始密码子后第 5~6 个碱基，使其由亮氨酸变为丙氨酸，从而构造一个适合细菌的表达结构，形成 C 端融合了长春花来源的 F3',5'H-P450 嵌合羟化酶，该酶能更好地催化槲皮素的形成（Leonard et al.，2006）。除对个别代谢途径中的酶进行修改外，将多种酶融合表达也是一种有效提高微生物生产花色苷产量的方法。研究表明，将拟南芥 F3GT 与牵牛花 ANS 的 N 端通过五肽连接进行融合，与单独的 ANS 或 F3GT 相比，2 种酶融合后可极大提高矢车菊素-3-O-葡萄糖苷的产量。与独立的酶相比，这种复合酶能够更有效地进行连续生化反应，催化产生大量中间产物，并迅速将其加至 F3GT 进行糖基化，从而减少了降解的可能，大大提高了花色苷的合成效率（Yan et al.，2008）。

（二）辅因子和辅底物对花色苷产量的影响

充足的辅因子和辅底物对花色苷生物合成中的电子转移、酶活化以及稳定性有重要意义。亚铁离子、抗坏血酸钠和 2-酮戊二酸分别是 ANS 的辅因子和辅底物，帮助 ANS 氧化底物，促进花色苷的生成（Turnbull et al.，2004；Welford et al.，2005）。研究表明，抗坏血酸钠和亚铁离子的加入显著提高了儿茶素底物的消耗和矢车菊素-3-O-葡萄糖苷的产量（Yan et al.，2008；Leonard et al.，2006）。但在生产花色苷的过程中，辅底物 2-酮戊二酸的添加并不是必需的，这可能是由于 2-酮戊二酸是三羧酸循环（Tricarboxylic Acid Cycle，TCA）的中间产物，因此其供应丰富（Yan et al.，2008）。

鸟苷二磷酸葡萄糖（Uridine Diphosphate Glucose，UDPG）是一种重要的葡萄糖供体，能催化 C3 位糖基化反应，使不稳定的花色苷变得稳定，UDPG 的含量可影响糖基化花色苷的生产。常用的处理方法是提高其生物合成基因的表达或对其降解途径加以抑制，如在 E. coli 中上调 UDPG 合成代谢途径的关键基因（pyrE、pyrR、cmk、ndk、pgm、galU）表达或抑制 UDPG 消耗通路，这些改造导致矢车菊素-3-O-葡萄糖的产量较对照组提高了 20 倍以上（从 4 mg/L 增加至 97 mg/L）（Leonard et al.，2008；Yan et al.，2008）。另一项研究显示，pgm、galU、ANS 和 3GT 过表达可以使矢车菊素-3-O-葡萄糖产量增加 57.8%（Yan et al.，2008）。

S-腺苷甲硫氨酸也是花色苷生产时必不可少的辅底物，如芍药素 3-O-葡萄糖甲基花色苷的生产。通过 CRISPR 干扰介导的转录阻遏物 MetJ 沉默技术可提高 S-腺苷-L-蛋氨酸的可用性，利用 UDP-葡萄糖和 S-腺苷甲硫氨酸的双重改善技术可将 E. coli 中芍药花色苷产量提高到 56 mg/L（Cress et al.，

2017)。

（三）转运蛋白对花色苷产量的影响

转运蛋白通常分为3类：摄取泵、外排泵和调节剂（Piddock et al., 2006）。许多工程微生物产生的目标产物对宿主菌株有害，从而抑制了目标产物的产量。解决该问题的有效方法是利用外排泵将细胞质内的目标化合物转移到细胞外环境，使产物维持在适当的水平，从而达到减毒的作用。例如，为提高 E. coli 中生物燃料的产量，引入一系列外排泵，负责将细胞内产生的生物燃料排出细胞外（Dunlop 等，2014）。这种方法已被广泛用于多种生物系统中，包括花色苷的生物合成。Lim et al.（2015）选择4种外排泵 acrAB、tolC、aaeB 和 yadH，发现只有 yadH 过表达能将细胞外矢车菊素-3-O-葡萄糖苷产量提高15%，说明矢车菊素-3-O-葡萄糖苷可能是 yadH 外排泵的底物；过表达 acrAB 和 aaeB 后细胞外儿茶素和花色苷的含量与野生型细胞相当；此外，去除1种负责儿茶素的分泌外排泵可进一步提高矢车菊素-3-O-葡萄糖的产量。

除微生物自身运输的调节外，对植物运输调节系统的引进也可能是改善工程微生物中花色苷产生的可能途径。花色苷在其天然植物宿主中被特定的转运蛋白转运并积聚在液泡中（Yazaki, 2005）。在玉米中，花色苷的转运主要是利用液泡膜中存在的三磷酸腺苷（Adenosine Triphosphate, ATP）结合转运体 ZmMRP3 实现的（Goodman et al., 2004），通过 Bronze-2 编码的谷胱甘肽 S-转移酶实现花色苷在液泡中的沉积（Marrs et al., 1995）。在其他植物中，花色苷的运输也与 H^+ 梯度有着密切的关系。在拟南芥中，Testa12 基因编码1种属于 MATE 家族的二级转运蛋白类蛋白，可通过 H^+ 梯度逆向转运花色苷至液泡中，而缺乏这种基因突变体会大大降低花色苷在液泡中的积累（Debeaujon et al., 2001）。但这些植物中的转运机制目前并没有在微生物生产花色苷的工程中得以应用或尝试。

（四）花色苷微生物生产中培养条件的优化

花色苷的高度不稳定性在微生物生产中是一个棘手的问题。在植物中，花色苷合成后稳定的存储在液泡中，与植物不同，细菌细胞作为人工生产的宿主缺乏花色苷稳定机制。由于代谢工程细菌细胞正常生长 pH 值约为7，使得花色苷合成后极不稳定。为稳定微生物中花色苷的合成，提出了两步催化法：第1步，在 pH 值=7 的培养基中培养细胞，以维持细胞正常生长和外源酶表达；第2步，培养到一定阶段后将细胞转移到 pH 值=5 的新鲜培养基中，以减少花色苷降解。此外，也可以添加相应保护剂，如添加谷氨酸可在低 pH 值条件下维持细胞生存，该方法可使 E. coli 的矢车菊素-3-O-葡萄糖产量提高 15 倍左右（Yan et al., 2008）。

除 pH 值外，其他因素在花色苷生产中也发挥重要作用，如诱导时间、补料量、溶解氧和温度。Lim et al., (2015) 发现定期诱导时工程化 E. coli 中矢车菊素-3-O-葡萄糖苷的产量最高，并且 UDPG 和儿茶素与定期诱导结合后提高了花色苷的产量。溶解氧对花色苷微生物生产具有双重效应。一方面，氧是 ANS 发挥功能的关键因素；另一方面，氧会氧化花青素。因此，最佳溶解氧浓度是微生物生产花色苷的关键。在利用微生物生产圣草酚的研究中发现，通过增加溶解氧浓度能促进儿茶素的合成，这可能与充足的还原型辅酶Ⅱ供应有关（Zhao et al., 2015）。然而，目前还没有关于氧气对花色苷产量影响的报道。温度是影响细胞活性和外源蛋白表达的关键因素。温度的波动会影响蛋白折叠，进而影响某些蛋白质形成，间接影响微生物生产有用化合物。在 E. coli 中生产阿夫儿茶素时，感应温度为 20 ℃时其产量可达到 22.9 mg/L，而温度为 10 ℃时只能达到 6.1 mg/L（Jones et al., 2016）。

四、酮类及醌类衍生物

酮及醌类衍生物色素的种类较少。常见的酮类衍生物色素有红曲色素、姜黄素、甜菜红素等。红曲色素是红曲霉菌在代谢过程中产生的一类聚酮体化合物的混合物（朱佳丽等，2017），有红色系和黄色系两大类（王金字等，2010），由羰基、醚键等官能团、共轭双键结构（-C=C-C=C-）发挥出降血脂、抗氧化、抗炎等作用（玛合沙提·努尔江等，2023）。姜黄素是从姜科、天南星科中的一些植物根茎中提取的一种黄色二酮类化合物，现代医学研究发现其具有抗氧化、免疫调节、抗癌和预防阿尔茨海默病等作用（Nelson et al., 2017）。姜黄素易受到光、热、氧、酸、碱等因素降解，因此限制了其应用效果与使用范围（罗晓莉等，2023）。甜菜红素是水溶性含氮色素，其来源包括甜菜根、苋菜和红肉火龙果等（高婧宇等，2024）。研究表明，光、热、氧、和 pH 值变化等因素都会影响其稳定性（Abedi-Firoozjah et al., 2023）。醌类化合物是一类含有两个双键的六碳原子环状二酮结构的芳香族有机化合物，主要包括苯醌、萘醌、菲醌、蒽醌等类型。天然的醌类化合物多为橙色或橙红色结晶，少数数呈紫色，是中草药（王恒等，2023；王娜等，2023）、芦荟（杨常碧等，2022）、核桃（张唯等，2018）中重要的化学成分之一。果蔬褐变也和植物中多酚被多酚氧化酶氧化为醌类最终产生黑色素有关（马烁等，

2023)。

冷藏后褐变是影响采后果实品质和商品价值的关键因素之一。近年来，冷藏果实褐变问题越来越受到国内外研究者的关注，包括桃（Lurie et al., 2005; Jin et al., 2014）、荔枝（Liu et al., 2011; Jiang et al., 2006; Sivakumar et al., 2010）、芒果（Nunes et al., 2007; Li et al., 2014）、梨（李江阔等, 2009; Sheng et al., 2016）、枇杷（郑永华等, 2000）和香蕉（Li et al., 2016）等水果。褐变分为酶促褐变和非酶促褐变，前者是果实褐变的主要原因。酚类底物、多酚氧化酶（Polyphenol oxidase, PPO）和氧气是酶促褐变的3个要素（Veltman et al., 1999; Leja et al., 2003; Nguyen et al., 2003）。通常情况下，酚类底物主要位于液泡中，PPO位于细胞质（如质粒或者线粒体）和细胞质膜中，二者因为位于不同的细胞器中而无法接触（郝利平等, 1998; Amaki et al., 2011; Lin et al., 2016），低温贮藏使果实内活性氧（Reactive oxygen species, ROS）过剩积累，发生膜脂过氧化，细胞膜的完整性被破坏，酚类底物和PPO的区室化分布被打破，PPO将酚类底物氧化成高活性的醌类物质，聚合之后形成褐色物质，从而出现褐变表型（Mathew et al., 1971; Mayer, 1987; Sheng et al., 2016）。因此，低温条件下果实细胞膜结构破坏被认为是褐变的主要原因。Conn（1984）认为，代谢产物的区室化分布能将有毒产物限制在固定的区域内，从而避免了对植物造成进一步的伤害，是植物自我保护的一种重要机制。

贮藏环境如温度、酸碱度、光照等因素均可对以上几种天然色素产生影响，对采后果蔬贮藏过程中色泽变化至关重要，因此，果蔬在采后贮藏过程中应注意环境条件变化，严格把控贮藏环境。除了色泽品质外，果蔬在贮藏过程中风味品质也会发生变化。

第二节 风味

一、糖

风味品质是衡量果实品质的重要指标之一，果实在采后贮藏过程中极易发生风味劣变，不仅影响品质，而且严重影响其市场价值。果实风味是由味感和嗅感构成，前者以甜酸味为主体，与糖、酸的种类和含量有关，糖酸比是风味的基础，后者取决于挥发性芳香物的种类和含量，而挥发性化合物的复杂性及其相互间的特定组合赋予了各类果实独特的风味（张海英等, 2008; Tieman,

2017）。水果的风味品质对消费者的选择具有重要影响。糖及其衍生物糖醇类物质是构成果实甜味的主要物质，蔗糖、果糖和葡萄糖是果实中主要的糖类物质。而酸味物质主要是有机酸，如苹果酸、柠檬酸、酒石酸等（陈敬鑫等，2021）。对于果实来说，可溶性固形物含量（Total Soluble Solids, TSS）是衡量水果品质的一个重要参数，TSS组成成分中可溶性糖所占比例最大。一般情况下，在果实成熟过程中，其糖分和TSS均会显著增长，而淀粉含量、酸度则降低。作为呼吸基质的一部分，TSS含量在果实采后逐渐下降（张海新等，2010）。在贮藏过程中，果实后熟进程仍在继续，TSS含量在一定程度上受到贮藏条件与时间的影响（苏青青，2014）。猕猴桃（徐昌杰等，1997）和李子（孙秀兰等，2001）果实采后可溶性糖含量呈先升后降趋势。果实收获后，组织细胞质和液泡中蔗糖的降解和合成是糖代谢和积累的主要途径。细胞质中的蔗糖通过中性转化酶（Neutral invertase, NI）转化为果糖和葡萄糖，或通过蔗糖合成酶（Sucrose synthase, SS）转化为果糖和脲苷二磷酸-葡萄糖；而果糖和葡萄糖通过果糖激酶和己糖激酶磷酸化为果糖-6-磷酸和葡萄-6-磷酸，果糖-6-磷酸进入糖酵解和TCA中，为其他代谢过程产生能量和中间体（Zhu et al., 2013）。一般来说，由于淀粉的分解，成熟果实的TSS会增加，这也与总糖和还原糖含量的增加有关（Teixeira et al., 2018）。在果实贮藏期间，细胞呼吸首先消耗有机酸，其次为糖类；只有当有机酸被过度消耗后，呼吸作用才开始利用糖类（陈敬鑫等，2021）。

二、酸

除了糖外，有机酸的类型和含量也会影响果实的风味。果实中的有机酸主要包括柠檬酸、苹果酸和酒石酸。通常，有机酸在果实发育初期积累，大多数果实的有机酸含量在果实采后成熟贮藏前期呈上升趋势，后期随着后熟的进行呈现下降趋势，这主要与其作为呼吸基质而被消耗有关。果实成熟期间葡萄糖异生作用可将部分有机酸转化为糖，而柠檬酸和苹果酸等有机酸可被呼吸作用消耗（Brizzolara et al., 2020）。但有些果实如香蕉（张明晶等，2002）和菠萝（屈红霞等，2001）则与上述情况正好相反。有机酸盐（酯）被认为是TCA循环的中间产物，其合成和降解与TCA循环密切相关（Liu et al., 2016）。可滴定酸（Titratable Acid, TA）包括果实组织中游离酸或与阳离子结合的酸，是影响果实风味品质的重要因素之一。果实中大部分有机酸以有机酸盐的形式存在，其含量与TA直接相关（Liu et al., 2016）。酸类物质也会散发出强烈的酸败气味，对水果的品质产生负面影响。脂肪酸是果实中主要的挥发性酸类物质。高浓度的酸性物质会产生强烈的"刺激性酸奶酪"气味，主要包括乙酸、己酸、辛酸、丁酸等（Zhao et al., 2024）。在果实腐败过程中，存在氧化发酵反应，乙醇被 ADH 氧化为乙醛，乙醛进一步被乙醛脱氢酶（ALDH）氧化为乙酸。乙酸具有明显的气味特征，果实劣变后其风味显著增强。该过程主要由醋酸杆菌和木糖驹形氏杆菌挥发所致（Lynch et al., 2019）。高浓度的脂肪酸导致番茄果实具有强烈的"酸奶酪"气味，其中己酸（酸、脂、干酪味）、辛酸（脂肪、蜡质、酸败、干酪味）和丁酸（刺激的酸味）是造成番茄果实异味的主要原因（Zhang et al., 2023）。在猕猴桃中检测到的有机酸类型已报道 83 种，但物种丰度不足，还有未知的有机酸种成分尚未鉴定（Tian et al., 2021；Choi et al., 2022）。

三、挥发性物质

香气是果实的重要品质之一，很大程度上影响着消费者的购买欲望（Defilippi et al., 2009）。果实香气成分是由多种挥发性化合物组成的复杂混合物，果实挥发性芳香物质依据人们的感官评价，可分为果香型、清香型、醛香型、甜香型及花香型等（张梅，2007）。依据化学结构不同可分为酯类、醇类、醛类、萜烯类及含硫化合物等，是果实风味形成的基本要素之一（Christian et al., 2000；Lara et al., 2003；Hadi et al., 2013）。香气化合物生物合成途径中的一个重要步骤是获得主要前体底物，目前，研究人员对果实挥发性香气物质的合成途径已经明确，即以脂肪酸、氨基酸和糖为前体物质，经过一系列酶促反应衍生形成。依据参与反应的前体物质不同，果实香气物质合成途径可分为脂肪酸代谢途径、氨基酸代谢途径及萜烯类代谢途径 3 种（Song et al., 2003）。果实香气成分中直链脂肪族醇、醛和酯类物质主要来源于脂肪酸氧化，主要合成途径是果实中的脂肪酸经 β-氧化产生酮酸和酰基辅酶 A（Acyl-CoA，CoA），进一步被还原为醛和醇，或者脂肪酸直接被脂氧合酶（Lipoxygenase，LOX）氧化，形成 C_6 醛和相应的醇类（张海英等，2008）。

果实在采后贮藏过程中，由于受到成熟度、遗传特性、贮藏和环境条件如氧气、温度、微生物、采后预处理等多种因素的影响，导致风味成分发生变化，包括风味降解或产生醛类、醇类、酯类、含硫化合物、酮类、酸类等异味化合物，也会发生腐烂，水果的风味品质可能会降低甚至使其不可食用，导致水果风味劣变（Obenland & Arpaia, 2019；Shi et al., 2019；Hadi et al., 2013）。果实风味劣变的发生可能导致异味的产生，包括乙醇、乙醛、乙酸乙酯等代谢化合物。此外，果实风味劣变会引发挥发性化合物组成和含量的变化。异味物质主要是指不属于水果本身的物质，或经过一系列生化反应而成为具有异味的化合物。特别是挥发性化合物可以在一定的气味阈值下散发出令人愉悦的香气，丰富了产品的风味特征。但是，一旦超过一定的气味阈值，可能会产生"青草味""辛辣味"，甚至"脂肪味""苦味"等气味特征，从而破坏产品的整体感官风味（Leonard et al., 2022）。例如，乙醛是一种具有"刺激性"感官特征的醛类物质，特别是在呼吸跃变型的果实中（Ceccarelli et al., 2020）。高浓度的乙酸乙酯具有"刺激性"和"涩味"的感官特性，严重影响果实原有的风味品质（Ceccarelli et al., 2020）。风味物质的组成和含量变化可能是由于果实原有特征香气物质发生降解，导致果实特征香气丧失造成的。或者可能由于某些非关键香气物质含量增加，将其转化为关键香气物质，从而引起果实香气的变化和果实风味劣变的发生（Zhao et al., 2024）。

（一）醇类物质

在早期的研究中，醇是气味感知的原始来源，醇类对果实香味形成起着重要作用。如乙醇是在水果中发现的一种常见的气味醇，较低浓度的乙醇有利于果实品质的保持并促进风味的形成，而高浓度则会产生异味（Pesis., 2005）。乙醇的发酵代谢不仅诱导乙醇大量积累，导致果实携带醇香气味，还诱导乙醇衍生酯的形成（Goldenberg et al., 2016）。乙醇脱氢酶（Alcohol Dehydrogenase，ADH）是生物体内短链醇代谢的关键酶，催化乙醇与乙醛可逆转换，在很多生理过程中起着重要作用（何双等，2019）。乙醇发酵是丙酮酸在丙酮酸脱羧酶（Pyruvate Decarboxylase，PDC）和ADH催化下重新转化为乙醇的两步过程（Goldenberg et al., 2016）。但乙醇也是果实厌氧代谢的产物，厌氧代谢的调控主要是通过PDC酶催化的，而ADH酶对其调控影响很小（Zhao et al., 2024）。除此之外，醇类物质还可能散发出不属于水果风味的"霉味、蘑菇味、脂味、油味、苦味、焦味"等气味，是水果风味劣变的直接反映。例如，1-辛烯-3-醇是一种具有"霉味"和"蘑菇味"气味的化合物，通常在腐烂的水果中容易被察觉。

（二）醛类物质

醛类物质是产生一类类似绿草香气的物质。对于呼吸跃变型果实来说，跃变前果实中存在的醛类物质是绿色/未熟果香味的重要来源；当成熟发生时，挥发性的醛类转变为酯类，以形成果实特有的果香味（Fellman, 2000）。乙醛是一种具有"刺激性"感官特征的醛类物质，是几乎所有果实的天然香气成分，尤其在苹果、猕猴桃、桃等水果中含量较高（Ceccarelli et al., 2020；

Thewes et al., 2019)。组成醛基的羰基（C = O）和氢原子（-H）决定了醛类物质独特的风味性质，乙醛也是水果风味劣变的特征指标，其在果实成熟过程中，甚至在有氧条件下也会积累（Pesis，2005）。果实中的乙醛是糖酵解过程中丙酮酸经 PDC 的脱羧作用形成的（Leonard et al.，2022）。因此，PDC 在乙醛形成的代谢机制中起着关键作用，而乙醛的形成对果实的各种生理和生态功能至关重要。在 ADH 和乙醛脱氢酶作用下，乙醛可直接转化为乙醇（Contreras et al.，2015；Belay et al.，2018）。在果实风味劣变中起重要作用的特征性醛类有很多，如癸醛、（E）-2-癸烯醛、（E）-2-庚烯醛、（E）-2-辛烯醛、庚醛、辛醛、壬醛等。研究表明，一些醛类如 2-甲基丙醛、糠醛、庚醛和癸醛生成量的下降可能与呼吸代谢有关（Belay et al.，2018）。大多数醛类物质是通过不饱和脂肪酸分解产生的，癸醛主要是由脂肪氧化酶对油酸的催化作用产生的。因此，这些特征醛类物质的变化是决定果实风味品质的重要因素之一（Zhao et al.，2024）。

（三）酯类物质

酯类物质是在果实成熟过程中由醇类和脂肪酸结合产生的化合物，酯类物质是形成果实香气的主要成分，但酯类物质的增加可能导致"过熟水果味"气味。在果实采后能量代谢过程中，丙酮酸在无氧呼吸的作用下经乙醇发酵形成乙酸乙酯，其含量的增加（超过一定的阈值）是果实风味劣变的重要影响因素之一。此外，乙醇可以与其他酰基 CoA 醇一起为 AATs 催化的酯化反应提供底物，导致挥发性乙酯的积累。例如，在贮藏过程中，一些酯类物质如乙酸乙酯、丙酸乙酯、2-甲基丙酸乙酯和 2-甲基丁烯酸乙酯的含量增加可能会影响果实的风味（Goldenberg et al.，2016）。其中乙酸乙酯和丙酸乙酯已被研究证实为赋予水果"过熟"风味的特征酯，是果实厌氧代谢的异味指示物（Thewes et al.，2018）。在对贮藏后的"摩罗"宽皮柑橘的鉴定研究中发现，"绿色"和"新鲜"气味的丧失，伴随着"过熟"和各种令人不愉快的气味，导致感官品质下降，果实风味变劣（Tietel et al.，2011）。因此酯类物质是果实贮藏过程中"过熟"异味的主要来源。

成熟果实所产生的酯类可大致分为直链型和支链型。直链酯类是由脂肪酸通过 LOX 途径和 β-氧化途径所产生的（Altisent et al.，2009），LOX 作为酯类香气成分合成途径中的关键限速酶，其在果实香气形成过程中起到重要作用（Altisent et al.，2009；Villatoro et al.，2008；Harb et al.，2008；Raffo et al.，2009）。LOX 途径以亚油酸和亚麻酸为底物，通过 LOX 催化，氧化形成氧羟基脂肪酸，再经过氢过氧化物裂解酶催化，还原为己醛或者己烯醛，之后经

ADH 催化，还原为相应的 C_6 或 C_9 醇。脂肪酸代谢途径是大多数植物果实中的 C_6 和 C_9 醇醛类物质和对应的酯类物质产生的主要来源（席万鹏等，2013；Chen et al.，2016；Tang et al.，2015）。脂肪酸可通过 β-氧化途径合成酰基 CoA，经过多步反应转化为内酯类香气物质（曹香梅，2019）。醇酰基转移酶（Alcohol Acetyl Transferase，AAT）是多功能蛋白，主要负责催化生物体内各种酰基化和去酰基化反应，在基因表达、代谢和信号传导中具有重要作用（何双等，2019）。醇类物质在酰基 CoA 的存在下经 AAT 催化，酯化形成酯类物质（姚苗苗，2018）。据报道，AAT 活性随着果实的成熟而提高，但其在贮藏过程中可被低氧环境所抑制，进而减少乙酸酯的生成量（Chervin et al.，2000）。

支链酯是由氨基酸代谢途径所产生的（Altisent et al.，2009；李秋棉等，2012）。果实中含有大量的游离氨基酸，经羧肽酶催化裂解成单个的游离态氨基酸（姚苗苗，2018），游离的氨基酸首先在脱羧酶作用下转变为胺，或者经支链氨基酸转氨酶等转氨酶的作用转化为支链酮酸，之后通过 α-酮酸脱羧酶等酶脱羧或者 ADH 脱氢反应生成支链醇和酰基 CoA，最后在酰基转移酶作用下，生成支链酯类物质。同时，酰基 CoA 作为酯类合成过程中的酰基供体，经过多步反应可形成酯类香气物质（李晓晶，2023）。现已证明参与氨基酸代谢途径的氨基酸包括亮氨酸、异亮氨酸、天门冬氨酸等（Garcia-Cayuela et al.，2012；Li et al.，2016）。部分果实的酯香型、果香型特征香气通过本途径产生，如番木瓜和草莓（Yuan et al.，2017）。

（四）萜类物质

萜类化合物是主要以单糖、糖苷为前体，经过催化反应生成的植物中种类最多的一类次生代谢物。该途径中，两分子乙酰 CoA 经酶促反应生成异戊烯基焦磷酸，在异构酶作用下生成焦磷酸 2-异戊烯酯，通过不同酶的催化作用生成焦磷酸香叶酯、焦磷酸香叶基香叶酯及法尼基焦磷酸，最后这些物质进一步合成甾醇和多萜次生代谢物（Wei et al.，2017；Yi et al.，2016）。研究表明，萜类物质是甜瓜和黄瓜的重要香气物质（Gonda et al.，2010）。在正常大气压条件下，只要具有足够高的蒸汽压力，大多数的半萜（C5）、单萜（C10）、倍半萜（C15），甚至一些二萜（C20）均可大量释放到空气中（Hadi et al.，2013）。据报道，在成熟或贮藏过程中，果实中与糖基结合形成糖苷的香气组分可通过酶或化学作用释放出来，进而参与形成果实的香气（Harb et al.，2008）。

除了醇、醛、酯和萜类化合物外，其他类化合物如酮类、酸类、内酯类等

第一章 采后果实感官品质变化

也能参与形成果实香气（陈敬鑫等，2021）。据报道，脂质降解也会导致酮类和呋喃类挥发性分子的产生，从而散发出气味（Zhao et al.，2024）。

第三节 硬度

果实在成熟和采后表现出不同的质地变化，如软化、硬化、粉质、絮败和脆化等（Contador et al.，2015）。其中硬度是指果肉抗压力的强弱，采后果实硬度变化会直接影响到果实的品质，硬度降低则表现出软化的现象，因此硬度是衡量果实品质的一个重要指标（张海新等，2010）。果实软化现象是指采摘后的果实在运输、储藏等过程中，发生了诸多的生理变化和生化反应，如细胞壁结构和成分变化、相关基因表达变化等，最终导致果实硬度下降，果实发生软化。此外软化的果实极易受到外力作用而损伤，或被细菌、致病菌等微生物污染，从而影响果实的货架期。果实软化是一个复杂的生理过程，对于某些果实而言，软化是果实成熟的一个主要标志，也是肉质果实成熟过程中最显著和不可逆转的特征之一（Shi et al.，2022）。软化会影响果实的外观、质地和风味，过度软化是降低水果质量和影响经济效益损失的主要因素之一（Wang et al.，2018；Li et al.，2020）。预计到2050年，为满足预期的100亿人口提供健康的饮食需求，水果和蔬菜生产需要大约翻一番（Willett et al.，2019）。减少目前多达一半的采后水果损失将极大地促进该目标（Porat et al.，2018）。果实过度软化、对微生物感染的易感性和贮藏紊乱是供应链中果蔬损失的主要因素。果实软化决定了可供贮藏、运输、零售的时间长短，也决定了消费者在家中的食用时长。搬运工人和托运人员需要坚固、结实的水果，以免在搬运或运输过程中造成水果的损伤，而消费者则希望水果在食用时才变得柔软，因此，延长可接受的软化期可使其达到消费者想要的良好状态（Brummell et al.，2022）。

目前已对肉质果实（脆的：苹果等；软的：桃和浆果等）发育和成熟过程中的软化进行了广泛的研究，其中包括呼吸跃变型果实番茄和非呼吸跃变型果实草莓，以及其他一些重要的经济水果，如猕猴桃、香蕉、桃和苹果等（Li et al.，2010；Tucker et al.，2017）。一些水果如山竹、枇杷和梨在采后则会表现出表皮硬度增加的现象（Cai et al.，2006；Li et al.，2010；Kamdee et al.，2014）。研究表明，果实质地的变化是细胞壁结构变化的结果，而细胞壁结构变化过程是由细胞壁组分的分解代谢造成的，该代谢是由细胞壁成分（如果胶、纤维素和半纤维素）的降解和细胞壁代谢酶（多聚半乳糖醛酸酶，Polygalacturonase，PG；果胶甲酯酶，Pectin Methylesterase，PME；纤维素酶，Cellulase，Cx；β-半乳糖苷酶，β-galactosidase，β-gal 等）协同作用完成的（Rose et al.，1999；Guo et al.，2018）。

果实软化是果实成熟的重要过程，细胞壁中间片层的溶解、细胞间黏附的减少和细胞壁修饰酶的作用可使薄壁细胞壁减弱造成果实软化，细胞壁修饰包括半纤维素解聚、果胶增溶解聚和果胶侧链的中性糖损失，从而改变细胞壁组分的机械性能。此外，由于中间层的溶解，细胞间的粘附性降低，这些变化与膨压降低共同导致果实软化（Brummell，2006；Mercado et al.，2019）。一些果实的硬度是由次级细胞壁（Secondary Cell Wall，SCW）增厚和果肉细胞中木质素的积累引起的（Li X et al.，2010；Wang et al.，2018），也有研究认为是由于初级细胞壁（Primary Cell Wall，PCW）和中胶层（影响管壁强度和细胞间黏附）的修饰以及细胞膨压降低形成的。研究表明，果实内含物淀粉的降解，会引起细胞壁结构发生改变，果实硬度下降（Mo et al.，2008），特别是香蕉果实（Shiga et al.，2011）。果胶是果实细胞壁的重要多糖组分，在成熟过程中有较大变化，包括增溶、解聚和中性侧链损失等情况的发生。草莓和苹果果胶含量差异较大，果实在成熟时分别具有软和脆的质地，表明果胶是质地变化的关键因素（Dhall，2013）。在柑橘、葡萄柚和柠檬等肉质果实中果胶含量可占细胞壁质量的 60% 以上。在货架初期，果胶以原果胶的形式存在，随着果实的成熟软化，PG、PME 和 β-Gal 等酶之间相互协调作用，使原果胶中的酸溶性果胶向水溶性果胶不断转化（钱丽丽等，2023）。即果实发生软化现象的主要原因是在许多细胞壁代谢相关酶的作用下，起支撑作用的细胞壁的结构发生改变，细胞壁物质被降解，细胞的透过性有所提高，细胞液向外渗出，果实硬度有所下降（Lin et al.，2018）。果实软化是一把双刃剑，它产生了消费者珍视的可食用品质，但也降低了采后的可运输性、可储存性和保质期，从而增加了经济损失和浪费（Tucker et al.，2017；Wang et al.，2022）。SCW 的形成也会对水果质量产生负面影响，通常会导致质地更坚硬，甚至不可食用（Cai et al.，2006）。

多年来，随着基因组和转录组测序以及蛋白组学的发展，有关软化相关的细胞壁变化的生理生化和分子水平研究在各种肉质果实成熟和采后贮藏期间已被广泛报道（Fischer et al.，1991；Goulao et al.，2008；Payasi et al.，2009；Tucker et al.，2017；Wang et al.，2022），包括苹果、番茄、梨和荔枝的软化过程（Win et al.，2019；Smith et al.，2002；Li. et al.，2010；Yi et al.，2021）。对细胞壁水解酶及其编码基因和转录因子（Transcription Factor，TF）的研究成为果实质地研究的热点（Seymour et al.，2013；Giovannoni et al.，2017；Wang et al.，2022）。这导致了大量与质地相关的 TFs 被鉴定出来，这些 TFs 能调节 PCW 和 SCW 的结构和性质以及淀粉降解。基于转录组技术，鉴定了许多细胞壁代谢中与软化相关的关键基因。在樱桃果实中瞬时表达 *PavXTH14*、*PavXTH15* 和 *PavPG38* 后，果实硬度显著降低，半纤维素和果胶含量也发生了变化（Zhai et al.，2021）。

除细胞壁代谢外，植物激素合成途径也会影响果实软化，并且合成途径中关键酶的表达受转录因子调控，且越来越多证据表明，转录因子和植物激素共同调节果实软化过程中的细胞壁代谢。目前，许多与果实软化相关的转录因子以及植物激素合成基因已被鉴定并用于调控果实软化，因此探讨转录因子在果实软化过程中对相关代谢途径中的基因的表达调控作用也很重要。采后果实发生软化甚至腐烂现象会影响其营养价值以及商品价值，也给水果采后储藏、销售等带来不便（钱丽丽等，2023）。这些研究的主要目的是了解质地变化的遗传和激素调节，并在育种计划和其他生物技术应用中利用这些理论，以改善水果质量和减少浪费（Shi et al.，2022）。因此研究采后果实软化机制对提高果实品质有重大意义。

第四节　营养

果蔬为人类生命活动提供了丰富的营养、能量和生物活性物质，如氨基酸、蛋白质、不饱和脂肪、膳食纤维、维生素、胡萝卜素、矿物质、多酚、类黄酮、花青素等（Siyuan et al.，2018）。然而，绝大多数的果蔬采后呼吸代谢旺盛，水分、可溶性糖、有机酸、维生素等营养物质严重损失，加速贮藏品质劣变，严重破坏其食用品质和商品价值（Pannitteri et al.，2017；陈琪琪等，2023）。

维生素 C 是果实最主要的营养物质，人体所必需的维生素 C 有 90% 来自水果和蔬菜。果实采后贮藏过程中维生素 C 含量随着贮藏期的延长而不断下降，主要是因为维生素 C 在中性和碱性条件下很容易被氧化。研究发现果实维生素 C 含量与果实的鲜脆状态相关，枣果只要在贮藏期间保持鲜脆状态，维生素 C 含量就保持较高的水平，一旦褐变软化，维生素 C 含量急剧下降（及华等，2005）。

水分是影响果实新鲜度、脆度和口感的重要成分，与果实的风味品质密切相关。但果实采后水分容易蒸散，果实大量失水，降低了果实的品质。因此为了保持采后果实的品质，应尽可能减少水分蒸发，例如使用塑料薄膜包装，可以降低果实采后水分的损失（张海新等，2010）。

可溶性糖和有机酸除了是风味物质外，也是重要的营养物质。研究表明，甜玉米常温贮藏 4 d 后，TSS、维生素 C 和有机酸等营养成分显著降低（Liu et al.，2021）；常温贮藏 14 d 后，蔗糖浓度和总糖浓度也显著下降（Hong et al.，2021）。

第二章 采后果实生理生化变化

采后果蔬品质劣变是制约果蔬贮藏和售卖的关键。一般来说,品质发生变化通常是由于果蔬膜脂代谢、细胞壁代谢、抗氧化代谢、能量代谢等生理生化代谢出现问题引起的。

第一节 呼吸代谢

果蔬采后需消耗能量来维持自身的生命活动,保持鲜活的状态(高红豆等,2021)。因此呼吸作用是基本的生命活动,也是植物具有生命活动的重要标志。果蔬采后同化作用基本停止,呼吸作用成为新陈代谢的主导,影响和制约着果蔬的贮藏时间、品质变化和抗病能力。采后果蔬由于呼吸代谢会消耗碳水化合物产生大量的中间产物,为其他合成过程提供原材料,也为生命活动和理化反应提供能量和前体物质(李美玲等,2019),而较高的呼吸作用会加速营养物质的消耗,产生 ROS,加速细胞膜的氧化损伤,从而进一步加速采后果实的衰老(Li et al., 2017)。呼吸作用强弱与采后果实品质劣变程度显著相关,与货架期也有关联(Tan et al., 2021)。果实采后一般有两种不同的呼吸方式:有氧呼吸和无氧呼吸,二者都是采后果实内在的生理机能。正常情况下有氧呼吸正常进行,当环境中 O_2 浓度过低时,植物组织会进行无氧呼吸,产生乙醛和乙醇等发酵代谢产物,降低有氧呼吸比例,从而减少细胞的能量供应,干扰和破坏果实正常的生理活动,对采后果蔬贮藏期品质产生重要影响(Li et al., 2018)。因此果实采后贮藏的关键之一就是尽可能降低果实的呼吸强度但又不能引起无氧呼吸的发生(张海新等,2010)。糖酵解途径是呼吸代谢的首要环节,也是植物体内蛋白质、脂质和糖类等有机物氧化分解的主要途径,较高的糖酵解活动会加速底物消耗和果蔬衰老(Li et al., 2017)。糖酵解-TCA 和磷酸戊糖途径是植物体内重要的有氧呼吸途径,可为机体提供能量(Slaski et al., 1996)。根据果实采后呼吸变化规律的不同,可将果实分为两种类型：呼吸跃变型和非呼吸跃变型。许多研究表明，苹果、桃、李、杏等果实采后均出现较为明显的呼吸跃变，称为跃变型果实；而葡萄、柠檬、菠萝、柑橘等果实采后没有明显的呼吸跃变，则称为非跃变型果实。影响果实采后呼吸的因素有很多，如果实的种类、品种、成熟度、温度、湿度、气体成分、机械伤害、病虫害和植物生长调节剂等，都会影响到果实采后的呼吸生理变化（张海新等，2010）。

第二节 活性氧代谢

ROS 是生物有氧代谢过程中的一种副产品，严重的低温胁迫可能通过破坏淬灭氧的清除系统从而对细胞造成损害。ROS 的积累造成果实生理机能下降，抗性减弱，容易遭受病原微生物的入侵进而出现细胞破损，果肉褐变，品质恶化，严重影响了果实的采后保鲜（田世平，2013；Tian et al.，2013）。ROS 的产生还能引发脂质过氧化、蛋白质氧化、核酸损伤和酶失活，并能激活程序性细胞死亡（张梦如等，2014）。在植物组织中，ROS 类型主要包括超氧阴离子自由基（Superoxide Anion，$\cdot O_2^-$）、过氧化氢（Hydrogen Peroxide，H_2O_2）、羟自由基（Hydroxyl Radical，$\cdot OH$）和单线态氧（Singlet Oxygen，$^1O_2$），都具有较强毒性，可以破坏重要的细胞成分，如 DNA、RNA、蛋白质和脂质（Li et al.，2019）。由于 $^1O_2$ 主要存在于叶绿体内，所以在果实后熟研究中，一般只研究 $\cdot O_2^-$、H_2O_2、$\cdot OH$。果实中的氧自由基产生于呼吸作用中，线粒体呼吸链的电子漏是植物体内产生氧自由基的重要来源。ROS 伤害植物的机理之一是它能够启动膜脂过氧化或膜脂脱酯作用，从而破坏膜结构。膜脂过氧化就是自由基对类脂中不饱和脂肪酸攻击引发的一系列自由基反应。果实在成熟衰老过程中，不断产生 ROS，破坏膜结构和功能的完整性，从而引发膜脂过氧化。丙二醛（Malondialdehyde，MDA）是膜脂过氧化的主要产物之一，人们常以 MDA 含量作为判断膜脂过氧化的标志，其含量的大小与膜脂过氧化程度呈正相关。

许多研究表明，膜脂过氧化是引起果实衰老的一个重要原因。不同成熟度"赞皇"大枣贮藏期间，MDA 含量与内源清除酶活性变化情况相反，证明枣果实后熟衰老与膜脂过氧化的加剧有关（及华等，2004）。"新红星"苹果在后熟衰老过程中，果实 MDA 含量升高，导致膜透性增高，加速了果实的衰老（关军锋，1992）。另外 MDA 还可与蛋白质上的氨基酸或核酸形成 Shiff 碱，它们通过引发结构变化来影响物质的代谢活动。Lurie et al.

（1991）认为，果实后熟的根本原因在于 ROS 代谢加强，对细胞产生毒害作用，导致细胞膜结构的破坏，促进果实软化和腐烂。总之，凡是能够协调与 ROS 代谢相关酶的作用和减少果实 ROS 产生的处理，都可以延缓果实衰老，提高果实的贮藏品质。

正常条件下，果实内的自由基和自由基清除系统处于平衡状态，果实正常生长；当果实遭到逆境胁迫或衰老时，二者的平衡失调，导致自由基清除系统清除自由基的能力下降，引起果实衰老。植物体内的自由基清除系统包括两类：一类是非酶系统，主要利用抗氧化物质如 β-胡萝卜素、生育酚、谷胱甘肽、抗坏血酸、类黄酮、甘露醇、维生素 E、类胡萝卜素和细胞色素等使 ROS 自由基的生成和消除处于动态平衡状态（Jarisarapurin et al., 2019）；另一类是机体可以通过清除酶系统如超氧化物歧化酶（Superoxide Dismutase，SOD）、过氧化氢酶（Catalase，CAT）和抗坏血酸过氧化物酶（Ascorbate Peroxidase，APX）、谷胱甘肽过氧化物酶（Glutathione Peroxidase，GSH）和谷胱甘肽还原酶（Glutathione Reductase，GR）等保护酶系统赋予植物抗 ROS 的保护作用（Joyce et al., 2005）。SOD 是存在于植物细胞中最为重要的清除自由基的酶类之一。它能催化 $\cdot O_2^-$ 发生歧化反应，所产生的 H_2O_2 被 CAT 分解成 H_2O 和 O_2 从而解除氧自由基对细胞的毒害作用。研究表明，大多数果实采后 SOD 活性随着果实的成熟不断上升，而后 SOD 活性随着果实的不断衰老又逐渐下降。过氧化物酶（Peroxidase，POD）和 CAT 也是果实内清除 H_2O_2 的主要保护酶。POD 具有多种生理作用，在 H_2O_2 存在的条件下可催化多种底物如还原型谷胱甘肽、抗坏血酸、酚类、芳香胺等的氧化，从而减少内源 ROS 自由基清除剂的含量，促进与果实褐变有关的反应。总之，只有 SOD、CAT 和 POD 这 3 种保护酶协调一致，才能使果实内的 ROS 自由基维持在较低的水平，减少自由基对果实的毒害，延缓果实的衰老，从而延长果实的贮藏寿命（张海新等，2010）。

低温能减少抗氧化物质含量和降低抗氧化酶活性，不能及时清除过剩的 ROS，从而导致 ROS 过量积累，造成膜脂中饱和脂肪酸积累并诱导氧化应激反应，从而降低了应激条件下的抵抗力，对机体造成威胁（杜秀敏等，2001；Yao et al., 2018；Valenzuela et al., 2017；Shen et al., 2017）。因此，提高果蔬体内抗氧化物含量和增强抗氧化酶活性对于降低采后果蔬冷害发生具有重要意义。

第三节　能量代谢

能量代谢对机体的生命活动至关重要。能量对于果实贮藏期间细胞各项生理功能的维持具有极其重要的作用（Li et al.，2018），在植物正常生命活动中，细胞通常能合成足够能量来维持组织的正常代谢，但当植物处于胁迫条件或衰老时能量代谢受损，不能维持正常的呼吸代谢和物质转化等，而ATP合成能力降低是引起组织能量亏缺的重要原因。组织能量亏缺是引起采后果蔬衰老的关键因素（Jiang et al.，2007）。果蔬采后品质劣变和衰老现象与其能量状态和新陈代谢密切相关（Liu et al.，2015）。此外，果蔬采后由于能量不足可引起褐变这一采后衰老现象（Veltman et al.，2003）。果蔬采后处于衰老状态时，由于能量合成能力下降，细胞因能量供应不足而出现功能紊乱和代谢失调，导致细胞膜结构被破坏、膜系统功能丧失，从而造成不可逆的细胞损伤，进而加速采后果蔬的衰老进程（Lin et al.，2016）。能量供应不足会导致细胞功能紊乱和代谢失调、细胞膜被破坏、功能丧失，同时导致防御系统无法清除多余的ROS，进而加速果蔬衰老（李美玲等，2019），并且保持较高的能量水平是植物细胞抵御外源微生物侵染的关键。可见，能量水平的高低也是评价果蔬耐贮性的重要指标之一（陈熙等，2024）。线粒体是生物能量代谢的主要细胞器，其中的 H^+-ATPase、Ca^{2+}-ATPase、琥珀酸脱氢酶（Succinate Dehydrogenase，SDH）、细胞色素C氧化酶（Cytochrome C oxidase，CCO）是线粒体呼吸代谢的关键酶，其活力的高低与线粒体功能密切相关（陈熙等，2024）。

研究表明，外源ATP含量与低温诱导的南果梨果皮褐变情况有关（Wang et al.，2018b）。ATP的产生受一系列特定蛋白质调控（Wang et al.，2013），在这些蛋白质中，ATP合酶，NADH脱氢酶和液泡质子-无机焦磷酸酶是参与氧化磷酸化途径的关键酶（Wang et al.，2017a）。这些酶与能量代谢和果实细胞膜完整性的维持有关。

第四节　膜脂代谢

著者对采后低温贮藏南果梨果皮褐变的膜脂代谢变化机理进行了研究（孙华军，2020），因此该部分内容以冷藏对南果梨果皮褐变的影响为主。

一、膜脂代谢简介

　　细胞内各种细胞器膜和细胞膜统称为生物膜，生物膜是细胞的重要组成成分，而膜脂、糖类和蛋白质是生物膜的主要成分，其中膜脂所占比例最多，是生物膜的主要化学成分，磷脂又是膜脂的主要组成成分，所以磷脂双分子层构成了生物膜最基本的支架，能维持生物膜的通透性和流动性，对维持细胞的内稳态起重要作用（孙德兰等，2008），在信号传导、膜流动和细胞骨架组成中发挥关键作用（de Bruxelles et al.，2001）。植物细胞磷脂组分结构中存在磷脂酰乙酸（Phosphatidic Acid，PA）、磷脂酰胆碱（Phosphatidylcholine，PC）、磷脂酰乙醇胺（Phosphatidylethanolamine，PE）、磷脂酰甘油（Phosphatidylglycerol，PG）、磷脂酰肌醇（phosphatidylinositol，PI）和磷脂酰丝氨酸（Phosphatidylserine，PS），这些磷脂在磷脂酶的作用下可以发生水解和转移，从而改变细胞膜的结构和功能。当植物接收到胁迫信号时，细胞膜中的磷脂被降解，细胞膜的结构和完整性被破坏。细胞膜对环境温度较为敏感，温度变化会影响膜脂酰基链的流动性，当温度降低时，生物膜中的不饱和膜脂含量降低，导致膜的流动性下降。在冷害温度下，细胞膜发生相变，由液晶态转变为凝胶态，膜脂的脂肪酸链由无序变为有序排列，膜的外形和结构也发生改变，出现孔道或龟裂，因此膜透性增大，膜内可溶性物质和电解质向膜外渗漏，细胞内外的离子平衡被破坏，同时膜上结合酶活性降低，酶促反应失调，呼吸作用下降，能量供应减少，植物体内有毒物质不断产生并积累，Lyons（1973）认为植物发生冷害后的各种代谢变化是次生或者伴生的，而冷害的最初始反应发生在生物膜的类脂分子上。此外，Sevillano et al.（2009）认为细胞膜物理相变是冷害在分子水平的第一个影响。该结论早已在叶绿体膜、类囊体膜和质膜中被证实（Krause et al.，1988；Hincha et al.，1987）。生物膜结构是一个动态平衡系统。当遭受非生物胁迫如冷害和气体伤害时，膜脂过氧化产物积累，细胞膜结构被破坏，进而影响细胞膜的功能，其特征是饱和脂肪酸相对含量增加，不饱和脂肪酸相对含量减少，表现为膜通透性增大和膜脂过氧化严重（Scotticampos et al.，2014）。当受到短期低温胁迫时，植物细胞内的脂质去饱和酶会催化饱和脂肪酸转化为不饱和脂肪酸，从而提高膜的流动性来适应环境温度的下降，维持植物自身内稳态，保护代谢活动正常进行。当植物长期受低温胁迫时，其自身防御系统已经不能阻挡低温造成的伤害，从而引发生物膜系统结构和功能紊乱。据报道，磷脂酶 D（PLD）、LOX 和脂肪酶催化膜脂质过氧化并破坏磷脂双层结构，破坏细胞膜的结构完整性（Lin et al.，

2016；Yi et al.，2008；Sirikesorn et al.，2013）。其中，PLD 能催化磷脂水解，然后水解产物可以被脂肪酶催化降解为游离脂肪酸（FFA）（Sirikesorn et al.，2013；Jia et al.，2015）。与未褐变的龙眼和荔枝果实相比，褐变的龙眼和荔枝果实中 PLD，脂肪酶和 LOX 活性更高（Yi et al.，2008；Zhang et al.，2018）。

  磷脂酶 D（Phospholipase D，PLD）是磷脂代谢中的关键酶，可以水解多种磷脂，如 PC、PI、PE 及其衍生物，生成 PA 和三磷酸肌醇（IP3）（Lin et al.，2023），在细菌、真菌、动物和植物中都有检测到（Yi et al.，2008；Jia et al.，2015；Morris et al.，1996）。第一个真核 PLD cDNA 是于 1994 年在蓖麻中发现的（Wang et al.，1994）。随后，陆续在拟南芥（Qin et al.，1997）、水稻（Ueki et al.，1995；Morioka et al.，1997）、玉米（Ueki et al.，1995）、烟草（Lein et al.，2001）和番茄（Whitaker et al.，2001）中克隆出了 *PLD* 基因。*PLD* 的不同成员可以水解许多磷脂分子包括 PI、磷脂酰胆碱（PC）、PE、磷脂酰甘油（PG）和磷脂酰丝氨酸（PS）（Munnik et al.，1998）。生理学研究表明，PLD 参与多种植物生长发育过程，如种子萌发、植株生长、花粉管萌发和伸长及叶片衰老等（Wang，2002；Wang，2005）。低温胁迫后香蕉中 PLDα 亚家族成员表达上调，PLD 活性被激活，LOX 活性、MDA 含量和细胞膜透性也增加；膜脂中 PA 的比例增加；PE、PC、PI 含量均下降，细胞膜受到损伤（Liu et al.，2011）。

  磷脂不仅是细胞膜的主要成分，而且作为信号分子参与植物代谢的各种活动，PA 可以直接刺激抗氧化酶，激活脱落酸的表达来刺激植物抗病机制的响应（Igbavboa et al.，2002）。此外，磷脂附着在多种不饱和脂肪酸链上，与磷脂一起行使细胞膜的功能。植物中常见的脂肪酸包括豆蔻酸、棕榈酸、花生四烯酸、硬脂酸、油酸、月桂酸、亚油酸和亚麻酸等。当植物受到胁迫或衰老时，磷脂酶被激发水解不饱和脂肪酸，产生多不饱和脂肪酸；随之而来的是 LOX 对不饱和脂肪酸的氧化作用以及不饱和度指数的降低，从而导致细胞膜的流动性降低（Abousalham et al.，1997）。

二、冷藏南果梨膜脂代谢变化及果皮褐变的发生

（一）不同冷藏期南果梨在常温货架期果皮褐变的发生情况

  褐变指数和褐变率是评估果实褐变严重程度的两个重要指标。如图 2-1 所示，冷藏 60 d 的南果梨无论是在出库当天还是整个货架期均未发生果皮褐变现象，而冷藏 120 d 的果实在货架第 3 d 即表现出褐变症状，随着货架期的

延长,褐变率和褐变指数均逐渐升高,至货架第 6 d 时,超过半数的果实有褐变症状,货架第 9 d,褐变指数剧增。

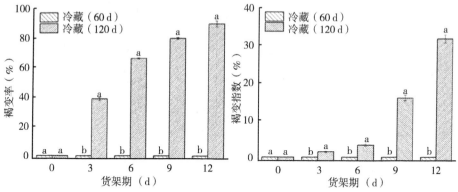

图 2-1 不同冷藏期南果梨常温货架期果皮褐变率和褐变指数变化

(二) 褐变果实果皮细胞的透射电镜观察

为了了解冷藏对南果梨果皮细胞结构的影响,本试验采用透射电镜观察了冷藏不同时间的南果梨在货架第 6 d 时果皮细胞的超微结构,结果如图 2-2 所示。短期冷藏南果梨的果皮细胞中拥有丰富和结构完整的细胞器,细胞膜结构完整且紧密贴近细胞壁,二者之间的界限清晰可见。然而,冷藏 120 d 的南果梨果皮细胞结构遭到破坏,细胞壁与细胞膜之间的界限变得模糊,叶绿体变得臌胀,叶绿体膜降解,且质体小球数量远多于短期冷藏样品。这表明长期冷藏南果梨果皮细胞的细胞膜和叶绿体等细胞器结构已经受到严重破坏。

(三) 不同冷藏期南果梨在常温货架期果皮组织细胞膜透性和 MDA 含量的变化

相对电导率是衡量细胞膜透性的重要指标。为了了解冷藏对南果梨果皮细胞膜的影响,测定了冷藏不同时间的南果梨常温货架期的相对电导率的变化。总体来说,出库当天,两组不同冷藏时间样品的相对电导率没有显著差异(图 2-3)。但冷藏后二者的相对电导率均随着常温货架期的延长而增加,且冷藏 120 d 的南果梨果皮相对电导率在整个常温货架期均显著高于同期冷藏 60 d 的样品 ($P<0.05$)。以上结果表明,短期冷藏的果实并未发生冷害,而长期低温胁迫使细胞膜发生严重损伤。

MDA 是脂质过氧化的终产物,通常用来表示氧化损伤的严重程度。与相

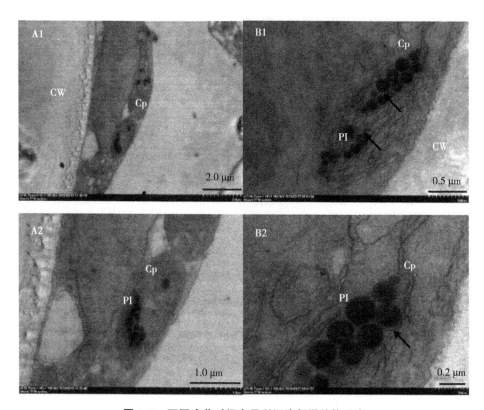

图 2-2　不同冷藏时间南果梨细胞超微结构观察

注：冷藏南果梨在常温货架第 6 d 的果皮组织细胞超微结构。A1，A2：冷藏 60 d 的果实；B1，B2：冷藏 120 d 的果实。CW：细胞壁；Cp：叶绿体；V：液泡；Pl：质体小球。(B1) 中的箭头表示类囊体和叶绿体膜的破坏。(B2) 中的箭头表示降解的细胞膜和质壁分离。

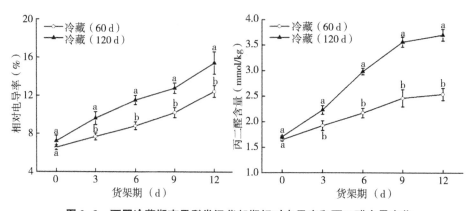

图 2-3　不同冷藏期南果梨常温货架期相对电导率和丙二醛含量变化

对电导率类似，出库当天，不同冷藏时间的果实 MDA 含量无显著差异。而在随后随着常温货架期的延长，不同冷藏期的样品中 MDA 含量均呈逐渐上升趋势，而且冷藏时间越长，上升幅度越大。

（四）不同冷藏期南果梨在常温货架期果皮组织脂肪酸组分及含量的变化

脂肪酸是生物膜的重要组成成分，在维持生物膜结构稳定性中起重要作用。本试验在南果梨果皮中共检测到了五种脂肪酸，其中包括 3 种不饱和脂肪酸如油酸（C18∶1），亚油酸（C18∶2）和亚麻酸（C18∶3），以及 2 种饱和脂肪酸如硬脂酸（C18∶0）和棕榈酸（C16∶0）。如图 2-4 所示，出库当天，不同贮藏期果实中五种脂肪酸相对含量均存在显著差异，其中，冷藏 120 d 的样品中三种不饱和脂肪酸相对含量均显著低于冷藏 60 d 的果实，而两种饱和脂肪酸相对含量恰恰相反，可见，低温贮藏过程中南果梨果皮细胞中各种脂肪酸的相对含量已发生了明显的变化。转入常温货架之后，两组样品中油酸相对含量均呈先升后降的变化趋势，二者的高峰均出现在货架第 9 d；亚油酸和亚麻酸相对含量的变化趋势相似，均呈先快速下降然后缓慢上升的趋势。尽管油

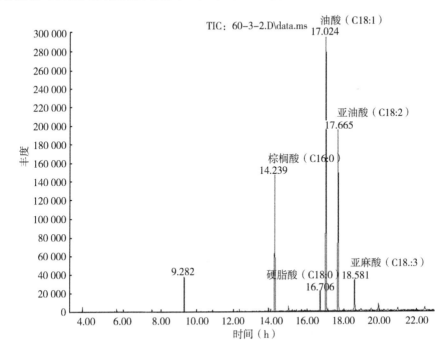

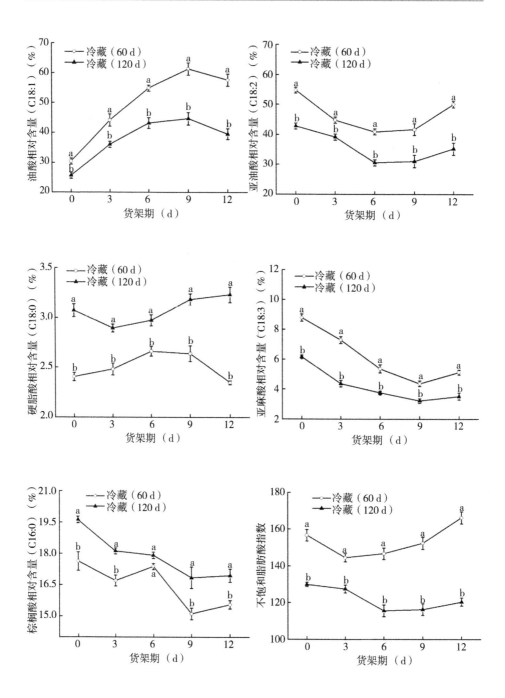

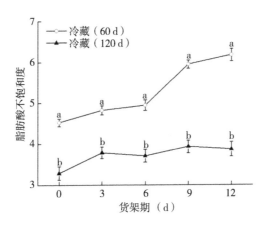

图 2-4　不同冷藏期南果梨常温货架期脂肪酸组成和含量变化

酸与亚油酸和亚麻酸相对含量的变化趋势有所不同，但是与短期冷藏相比，经过长期低温贮藏的果实中三种不饱和脂肪酸相对含量始终处于较低水平。与之相反，长期冷藏的果实中硬脂酸和棕榈酸的相对含量在整个货架期均显著高于短期贮藏的果实。进一步分析发现，无论是在出库当天还是整个货架期长期冷藏的果实不饱和脂肪酸指数和脂肪酸不饱和度都远低于短期贮藏的果实，由此可见，低温胁迫影响了南果梨果皮细胞脂肪酸的相对构成，降低了不饱和脂肪酸指数和不饱和度，而且这种影响不仅表现在冷藏后的常温货架期，在冷藏过程即可明显看到。

（五）不同冷藏期南果梨在常温货架期果皮组织膜脂组分及含量的变化

在冷藏后常温货架期第 6 d 检测到了六种磷脂包括磷脂酸（PA），磷脂酰胆碱（PC）、磷脂酰乙醇胺（PE）、磷脂酰肌醇（PI）、磷脂酰甘油（PG）和磷脂酰丝氨酸（PS），三种溶血磷脂包括溶血磷脂酰甘油（LPG），溶血磷脂酰胆碱（LPC）和溶血磷脂酰乙醇胺（LPE）以及两种半乳糖脂，包括单半乳糖基二酰基甘油（MGDG）和二半乳糖基二酰基甘油（DGDG）。如图 2-5 所示，冷藏 120 d 的南果梨总膜脂质含量显著低于冷藏 60 d 的样品。与短期冷藏样品相比，120 d 的南果梨 DGDG 含量降低了 28.6%，此外，长期冷藏的南果梨中 PA，PG，PC，PE，PG，LPC 和 MGDG 的含量明显低于短期冷藏的样品。长期冷藏的果实中 PG 和 DGDG 的百分比显著低于短期冷藏的样品。进一步分析了膜脂分子种类含量，从图 2-6 可以看出，经过长期冷藏后的南果梨

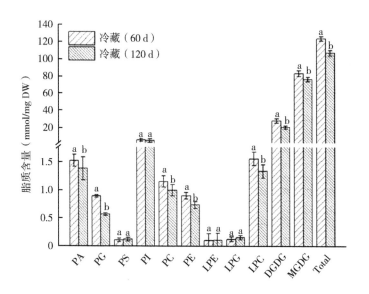

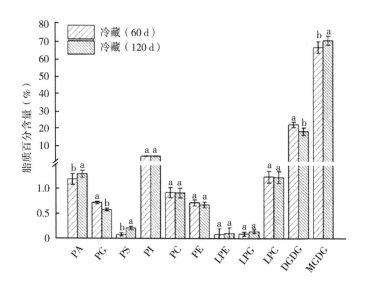

图 2-5　不同冷藏期南果梨膜脂组成和百分含量变化

果皮中溶血磷脂 18∶1 LPE 和 LPC 以及 16∶0 LPC 含量降低，其余溶血磷脂含量呈升高趋势，或变化不大。对于糖脂而言，36∶4、36∶6 和 34∶1 MGDG 是导致 MGDG 含量下降的主要脂质，而 DGDG 中大部分脂质经过 120 d 冷藏后均出现下降的变化趋势，其中 36∶4 DGDG 是主要贡献者（图 2-7）。对于 PA 分子，褐变的南果梨果皮中 36∶4 和 36∶3 PA 均显著高于未褐变果实，而褐变果实中 34∶1 PG、34∶2 PI、36∶2 PC，以及 36∶2 和 36∶3 PE 是下降的主要磷脂分子（图 2-8）。以上分析结果表明，脂类的组成和含量在长期冷藏南果梨中发生了显著变化。

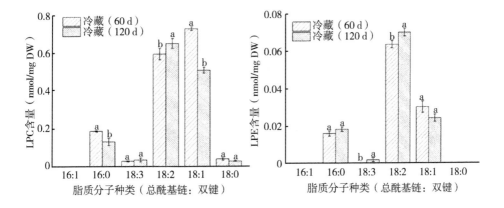

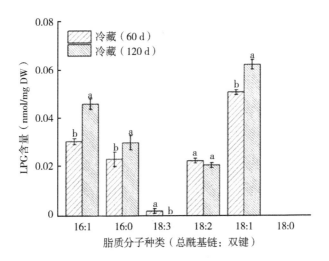

图 2-6　不同冷藏期南果梨溶血磷脂脂类分子含量变化

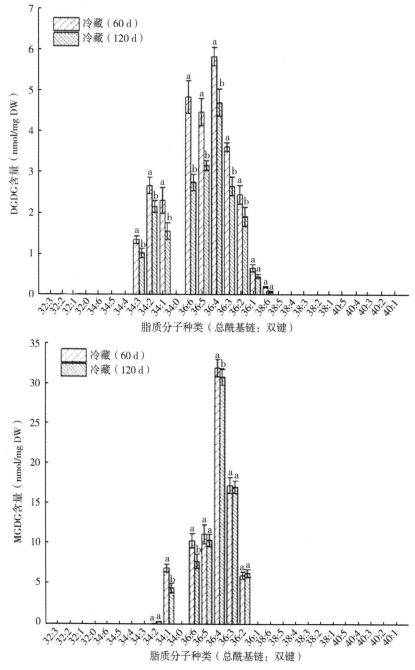

图 2-7 不同冷藏期南果梨糖脂脂类分子含量变化

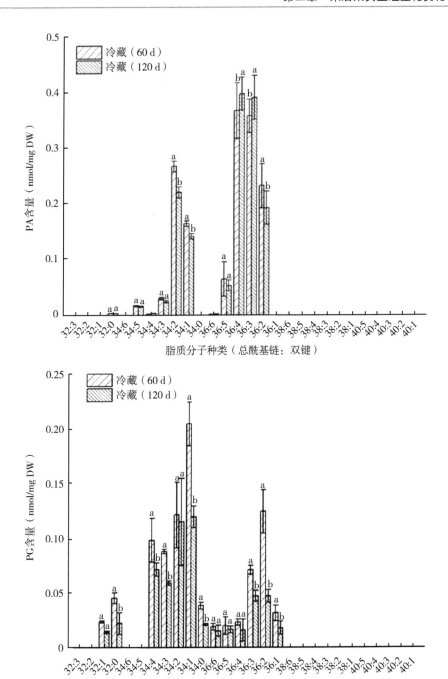

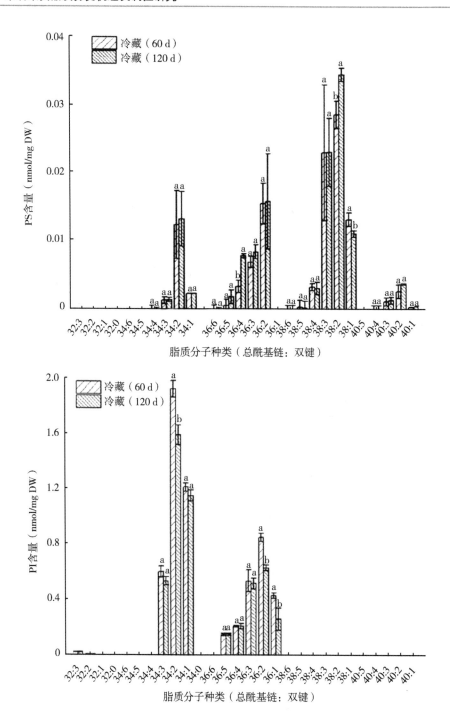

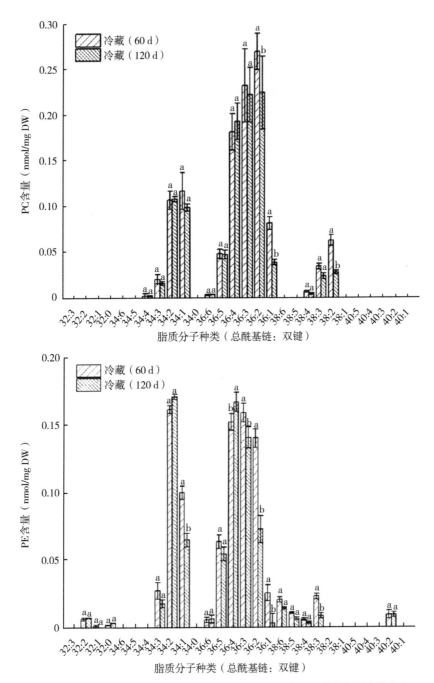

图 2-8 不同冷藏期南果梨 PA、PG、PS、PI、PC 和 PE 脂类分子含量变化

（六）不同冷藏期南果梨在常温货架期果皮组织膜脂代谢关键酶活性和基因表达的变化

1. 磷脂酶 D 活性及其基因表达变化分析

PLD 是膜脂代谢的关键酶。为了分析长期冷藏对不同 PLD 家族成员的影响，检测了冷藏不同时间的南果梨在常温货架期 PLD 活性和其家族成员的表达量变化，包括 PuPLDα1，PuPLDα4，PuPLDβ1，PuPLDδ 和 PuPLDζ2。出库当天，不同贮藏期果实中 PLD 活性和 PuPLDα1 转录水平没有显著差异，而冷藏 120 d 的样品中 PuPLDα4，PuPLDδ 和 PuPLDζ2 相对表达水平均显著低于冷藏 60 d 的果实，而 PuPLDβ1 相对表达水平恰恰相反，可见，低温贮藏过程中南果梨果皮细胞中 PLD 家族成员的转录水平已发生了明显的变化，但 PLD 活性没有显著差异。转入常温货架后，两组样品的 PLD 活性呈先升高后降低的变化趋势，二者均在货架第 9 d 时达到峰值（图 2-9）。与短期冷藏相比，经过长期低温贮藏的果实中 PLD 活性始终处于较高水平。与之类似，长期低温贮藏的果实中 PuPLDβ1 表达量在整个常温货架期显著高于同期短期冷藏果实。PuPLDδ 和 PuPLDζ2 的变化趋势相似，在常温货架期，其表达量快速下降，与长期贮藏果实交替变化。不同贮藏期的果实 PuPLDα1 表达量的差异主要表现在货架的中后期，表现为货架中期时长期贮藏果实的 PuPLDα1 表达量高于短期贮藏果实，而在货架后期两种冷藏时间的样品 PuPLDα1 表达量变化趋势恰恰相反。在货架前期和中期，长期冷藏 120 d 的果实 PuPLDα4 表达量低于贮藏 60 d 的果实，而在货架后期，长期贮藏果实的 PuPLDα4 表达量显著高于短期贮藏的果实，由此可见，低温胁迫诱导了南果梨果皮细胞 PLD 家族成员表达，从而使 PLD 活性增加，加速膜脂代谢进程，其中，PuPLDβ1 在应答长期低温胁迫更为明显。

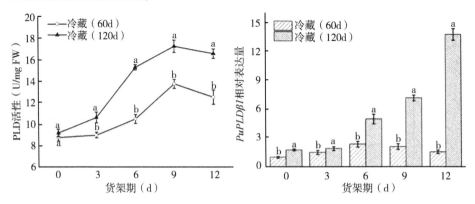

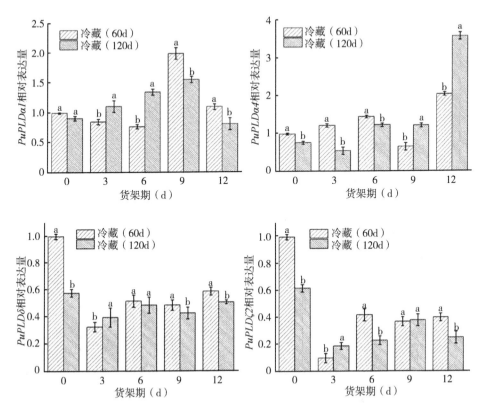

图 2-9　不同冷藏期南果梨常温货架期磷脂酶 D 活性及基因家族表达水平变化

2. 脂氧合酶活性及其基因表达变化分析

与 PLD 活性类似，出库当天，两种冷藏时间的果实 LOX 活性无显著差异，表明冷藏期间 LOX 活性变化不明显。在常温货架期，二者的 LOX 活性均呈先升高后降低的变化趋势，冷藏 60 d 的果实 LOX 活性在常温货架期第 6 d 时急剧升高，而冷藏 120 d 的果实在常温货架期第 3 d 时已经开始骤然升高，二者均在第 9 d 时达到峰值，然后骤降。然而，在整个常温货架期长期冷藏果实的 LOX 活性均显著高于短期冷藏果实。结果表明，长期冷藏能诱导 LOX 活性增加。如图 2-10 所示，冷藏过程中，两组果实的 PuLOX1d 和 PuLOX2S 相对表达量没有显著差异。与 LOX 活性类似，两种冷藏时间的果实 PuLOX1d 相对表达量均随着常温货架期的延长呈先上升后下降趋势，第 9 d 时达到峰值。此外，冷藏时间越长的果实 PuLOX1d 的相对表达量越高，且显著高于同期短期冷藏的果实。冷藏 60 d 的果实 PuLOX2S 相对表达量在常温货架期也呈先

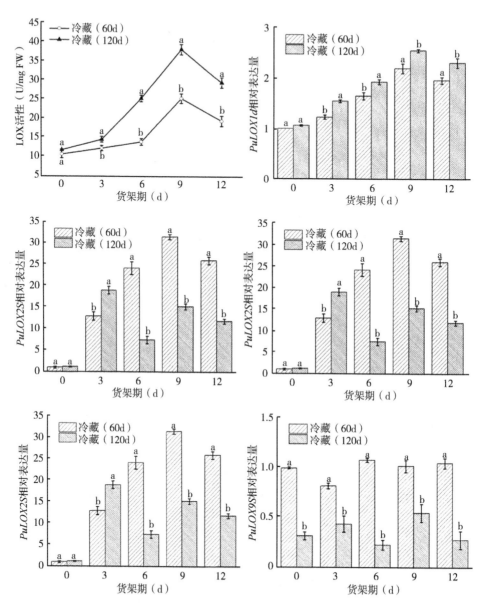

图 2-10 不同冷藏期南果梨常温货架期脂氧合酶活性及基因家族表达水平变化

升高后降低的变化趋势,第 9 d 时达到峰值。长期冷藏的果实 PuLOX2S 相对表达量在常温货架期第 3 d 时达到峰值,随后波动变化。在常温货架中后期,长期冷藏果实的 PuLOX2S 相对表达量显著低于短期冷藏果实。与 PuLOX1d 和

PuLOX2S 相对表达量不同，出库当天，长期冷藏果实的 PuLOX9S 相对表达量显著低于短期冷藏果实。在整个常温货架期，短期冷藏果实的 PuLOX9S 相对表达量变化不大，然而，长期冷藏果实的 PuLOX9S 相对表达量呈波动变化趋势，且显著低于同期短期冷藏果实。结果说明，长期低温胁迫能诱导 LOX 活性增加，促进脂质过氧化的发生。

3. 脂肪酶活性及其基因表达变化分析

如图 2-11 所示，与 PLD 和 LOX 活性变化不同，出库当天，冷藏 60 d 和 120 d 的果实脂肪酶活性存在显著差异。冷藏后，随着常温货架的延长，两种冷藏时间的果实脂肪酶活性均呈先降低后升高的变化趋势。短期冷藏果实在货架第 3 d 急速下降，至货架第 9 d 时达到最低值。长期冷藏在货架前 6 d 快速下降，随后骤然升高。此外，长期冷藏的果实脂肪酶活性在货架中后期显著高于短期冷藏果实。与脂肪酶活性类似，出库当天，两种不同冷藏时间的果实 Pulipase 相对表达量存在显著差异。冷藏 60 d 的果实 Pulipase 相对表达量在常温货架前 6 d 缓慢降低，至第 9 d 时快速降低，随后突然升高。然而，冷藏 120 d 的果实 Pulipase 相对表达量在常温货架前 3 d 基本没有变化，第 6 d 时突然升高并达到峰值，随后波动变化。同样，在常温货架中后期，长期冷藏果实的 Pulipase 相对表达量显著高于同期短期冷藏果实。结果表明，长期低温冷藏也能诱导脂肪酶活性增加，尤其在常温货架的中后期。

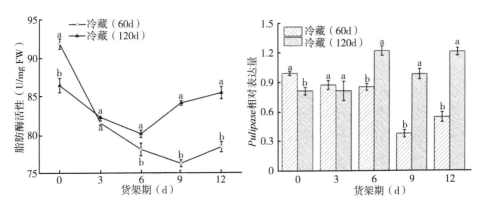

图 2-11 不同冷藏期南果梨常温货架期脂肪酶活性及基因家族表达水平变化

冷藏能延缓采后果蔬后熟衰老和防止腐烂，是有效延长采后果蔬贮藏期和维持果实品质的有效方法（Bourne，2006）。但是，长期冷藏能使果蔬发生冷害，表现为水渍状、表面组织凹陷、变色、褐变、香气变淡和丧失后熟能力等

（金鹏等，2012）。如长期冷藏能使南果梨发生果皮褐变和香气变淡的冷害症状，能使青椒发生水渍斑冷害症状等（Wang et al., 2017a; Kong et al., 2018）。褐变是由于在胁迫条件下，位于不同细胞器中的酚酶与酚类底物的区室化分布被打破，在有氧条件下二者接触，酚酶催化酚类物质氧化生成有色醌类物质，聚集之后形成褐变表型，因此，生物膜结构被破坏是冷藏南果梨果皮褐变的关键。在本章研究中，我们评估了不同冷藏时间南果梨的果皮褐变情况，并观察了褐变发生时南果梨果皮细胞超微结构，以便在细胞水平更好地了解长期低温胁迫对南果梨果皮造成的伤害。本试验发现短期冷藏南果梨无论在出库当天还是出库后的常温货架期均无褐变表型，而长期冷藏果实在货架第3 d时出现褐变现象，第6 d时褐变开始变严重，此时的细胞质膜发生降解，叶绿体轮廓模糊，叶绿体膜和类囊体膜也发生降解，质体小球数量增多，推测造成该结果的主要原因是类囊体膜中ROS过量积累，使得生物膜结构被破坏，南果梨果皮中的酚酶与酚类底物区室化分布被破坏，二者接触后产生褐变表型。这些结果揭示了长期低温胁迫对南果梨造成了严重损伤。

生物膜功能改变与膜脂过氧化密切相关。膜脂过氧化是指生物膜上的磷脂分子结构骨架不饱和脂肪酸中的自由基被氧化，生成过氧化物和MDA，从而破坏细胞膜（杨婷，2018）。细胞膜是第一个屏障，能将细胞与外环境分离，它也是环境胁迫下受伤的主要位置（Zhang et al., 2010）。植物细胞膜是由磷脂双分子层组成，正常情况下呈流动的液晶态。而细胞膜的甘油脂由流动凝胶相转变成晶体固态相的温度与脂肪酸的不饱和水平紧密相关。在低温胁迫下，植物的膜系会发生相变，由液晶态转变为凝胶态，使膜发生收缩，镶嵌在膜上的膜蛋白（酶）被固定，膜系统遭到破坏，膜上出现孔道或者龟裂，丧失了细胞膜的主动运输和选择透过特性，因此，膜透性增大，膜内大量电解质和可溶性物质向膜外渗透，破坏了细胞内外的离子平衡（赵金梅等，2009）。PLD，脂肪酶和LOX参与膜脂代谢，被认为是膜脂降解和过氧化过程中的关键酶（Katagiri et al., 2010; Sun et al., 2010）。在胁迫环境下，如低温、机械伤等，PLD介导的水解作用增加，导致磷脂降解为FFA，随后成为LOX的底物，LOX可以将FFA氧化为脂质过氧化物和ROS，导致膜损伤（Liu et al., 2011; Lin et al., 2017）。研究表明，有耐冷性的植物比冷敏感植物有更多的不饱和脂肪酸丰度，植物在适应低温环境过程中不饱和脂肪酸比例和去饱和酶活性都增加（Palta et al., 1993）。在冷藏期间，没食子酸丙酯处理过的龙眼果实脂肪酶、PLD和LOX活性明显降低，维持了更高的不饱和脂肪酸水平和不饱和脂肪酸指数和脂肪酸不饱和度，延迟了膜脂代谢并维持了膜的完整性，从而减轻了果皮褐变这一冷害症状（Lin et al.，2017）。同样，Yi et al.（2010）发现 ATP 处理过的荔枝果实中 PLD、脂肪酶和 LOX 活性显著增加低于未处理果实，且外源处理降低了果实的膜透性和 MDA 含量，表现出更低的果皮褐变现象。Wang et al.（2019）发现外源甜菜碱处理能抑制 PLD、LOX 和脂肪酶活性及相关基因表达，使不饱和脂肪酸和脂肪酸不饱和度更高，转录组结果显示脂肪酸生物合成和去饱和相关基因转录水平增强，而脂肪酸降解相关基因的转录水平减弱，因此，处理过的桃果实通过上调膜脂肪酸的生物合成和去饱和代谢来维持正常的细胞膜结构和功能，从而减轻桃果实的冷害。在本章中，冷藏 120 d 的南果梨发生了冷害，表现出果皮褐变的症状，而冷藏 60 d 的果实并未出现这一冷害症状，说明长期冷藏是导致果皮褐变的关键，该研究结果与前任研究结果一致（Sheng et al.，2016）；冷藏 120 d 的果实 PLD、LOX 和脂肪酶活性及相关基因表达显著高于同期短期冷藏果实，并且不饱和脂肪酸相对含量降低，饱和脂肪酸相对含量升高，脂肪酸不饱和度和不饱和指数均明显降低，伴随着相对电导率和 MDA 含量显著升高，表明长期冷藏刺激了膜脂代谢相关基因 PLD、LOX 和脂肪酶的表达，加速了膜脂降解和过氧化进程，从而导致果皮褐变的发生，其中，$PuPLD\beta1$、$PuLOX1d$ 和 $Pulipase$ 表达量都显著提高，是参与该过程的关键基因，但 PLD 直接催化膜脂降解，又在 PuLOX1d 和 Pulipase 的催化下促进下游反应进程，因此本研究后续对 $PuPLD\beta1$ 的转录调控机理进行了深入研究。Sheng et al.（2016）和 Wang et al.（2018b）的研究也发现冷藏 120 d 的南果梨在出库后常温货架期会出现果皮褐变的现象，短期冷藏的果实则没有出现，并且褐变果实的 PLD、LOX 和脂肪酶活性和基因表达显著高于未褐变果实，但是他们并未对基因家族成员的表达进行分析，本研究系统分析了 PLD 和 LOX 家族成员的表达，从而筛选出膜脂代谢关键基因，为后续深入研究膜脂代谢介导的褐变机理奠定重要基础。

膜透性和流动性的变化与遭受环境胁迫时膜脂质组成的变化有关。因此，我们分析了同期可变与非褐变南果梨的膜脂变化。检测到了六种磷脂，三种溶血磷脂和两种糖脂，结果显示，长期冷藏的果实总膜脂水平明显降低，这些变化可能由糖脂 DGDG 和 MGDG 的急剧降解引起的，它们分别占总膜脂的约 17.86% 和 67.06%。此外，我们发现长期冷藏还会降低 PC 和 PE 的含量，因此推测长期冷藏会加速这些脂质降解。PA 是一种单分子，由 PLD 水解其他结构性磷脂产生，参与各种生物过程，可响应冷胁迫（Welti et al.，2002）。据报道，PA 过量积累可能产生氢过氧化物和自由基，最终导致膜脂受损（Choudhury et al.，2017）。研究表明，敲除 $PLD\alpha1$ 基因后植物的抗冻能力增强，同时 PA 含量降低（Welti et al.，2002）。此外，溶血磷脂也是信号脂质，它来自膜脂质的水解，然后被释放到细胞外空间，在那里它们可以被细胞外受体识别并启动信号传导途径（Hou et al.，2016）。在极端温度、缺水和盐胁迫下，溶血磷脂的含量会急剧增加（Welti et al.，2002）。在青椒中，低温储存会增加 PA、LPC 和 LPE 的比重，导致冷害更为严重（Kong et al.，2018）。我们的研究结果表明，长期冷藏的南果梨中发 PA、LPE 和 LPG 的含量显著高于同期短期冷藏果实，这为冷胁迫下膜脂质重塑提供了证据，也为 PE 和 PG 的降解提供了依据。

第五节 细胞壁代谢

细胞壁作为植物抵御病原体攻击的第一道屏障，主要由初生壁、中间层、次生壁三层组成。初生壁由纤维素、半纤维素、果胶、糖蛋白等组成，次生壁包含以木质化为特征的木质素、纤维素等（Lampugnani et al.，2018），中间层由果胶组成，负责将相邻的细胞粘连在一起。有研究表明，木葡聚糖在有限的位置与纤维素微纤相互作用，而果胶与纤维素微纤广泛结合，大约一半的纤维素表面被果胶覆盖（Cosgrove，2014；Park et al.，2012，2015）。细胞壁的组成决定了细胞的结构，同时影响了细胞的大小、形状和功能（Zhang et al.，2021）。一般果实的细胞壁结构由薄壁层组成，尤其是成熟后的肉质果实中只有初生壁和中间层（Yokoyama et al.，2014）。因此当果实发生软化现象时，在细胞壁代谢相关酶的作用下，细胞壁中的结构多糖被降解，使得细胞壁结构和物质成分发生改变。引起细胞壁代谢的酶有很多，如 PME、PG、纤维素酶、β-半乳糖苷酶（β-galactosidase，β-gal）等（Lin et al.，2019）。在不同生长阶段的相同果实以及不同种类的果实中，引起果实软化的关键酶有所差异。Ben et al.（1993）研究了牛油果果实软化的主要原因，结果表明，PG 会导致牛油果果胶成分发生改变，使原果胶降解为可溶性果胶，而可溶性果胶能溶于水（庞荣丽等，2012），导致细胞壁结构松动和降解，因此牛油果果实硬度有所下降。Brummell et al.（2004）研究了猕猴桃果实软化原因，结果表明，纤维素酶会造成猕猴桃中的纤维素发生水解，导致细胞壁解离，果实发生软化。

一、果胶及果胶酶

果胶是一类与果实软化相关的多糖，它主要由半乳糖醛酸、均半乳糖醛酸、鼠李糖半乳糖醛酸-Ⅰ、鼠李糖半乳糖醛酸-Ⅱ、木糖半乳糖醛酸组成

（Pose et al.，2019）。当果实发生软化现象时，细胞壁中的原果胶降解为可溶性果胶，细胞间粘连力下降导致细胞壁结构松散，其中起到作用的酶有 PG 以及 PME。PME 是直接作用于果实细胞壁中果胶成分的主要酶之一，PME 能使半乳糖醛酸残基之间的酯键发生断裂，生成半乳糖醛酸和甲醇，果胶变成低甲酯化果胶，从而有助于 PG 对果胶的水解，在梨（刘剑锋等，2004）、鹰嘴蜜桃（汤梅等，2018）、香蕉（Ali et al.，2004）等研究中均已证实该结论。Hu et al.（2012）研究了用 NaHS 处理对草莓果实软化的影响，研究结果表明，NaHS 处理能抑制 PME 活性从而有效延缓草莓果实软化。目前，多数研究表明，PME 有利于 PG 的反应，促进果胶的降解（Zhang et al.，2019）。PG 是作用于果胶成分的另一种关键酶，PG 能使半乳糖醛酸主链上的 1,4-α-D-半乳糖苷键发生断裂，生成半乳糖醛酸和寡聚半乳糖醛酸，达到破坏果胶结构的目的，使细胞壁结构发生改变，果实发生软化现象。PG 是导致果实发生软化现象的关键因素，这在李（Sharma et al.，2012）、猕猴桃（李圆圆等，2018）、枣（李欢等，2017）、苹果（张娟等，2015）等果实软化研究中均被证实。Zhong et al.（2007）研究了 1-甲基环丙烯（1-Methylcyclopropene，1-MCP）复合壳聚糖涂膜处理对台湾青枣的影响，结果表明，该处理能有效抑制 PG 的活性，达到延长货架期的目的。在桃、葡萄、香蕉等其他肉质水果软化的研究中同样发现果实软化也受到 PG 基因表达的调控（Qian et al.，2021；Khan et al.，2019）。

二、纤维素及纤维素酶

纤维素作为植物细胞壁组成成分对细胞起着重要的作用。当果实发生软化现象时，纤维素酶活性逐渐增加，细胞壁的骨架物质纤维素被降解，导致细胞壁结构松散。纤维素酶能使 β-1,4-糖苷键断裂，生成葡萄糖。罗自生（2005）研究了柿果实的软化过程，结果表明，在软化过程中纤维素酶活性迅速提高，使纤维素发生降解，导致果实软化。有学者研究桃果实软化过程，研究发现纤维素酶活性的提高使桃果实的细胞壁结构发生改变，加速桃果实的软化（Brummell，2006）。赵云峰等（2012）发现，茄子采后贮藏过程中，纤维素酶活性不断提高，且与果实硬度之间呈显著负相关。但在京白梨软化过程研究中发现，当京白梨果实硬度降低、纤维素含量下降时，纤维素酶活性也逐渐降低，因此推测纤维素酶与京白梨果实软化不相关（魏建梅等，2009）。对其进行相关性分析，发现纤维素与 PG、β-半乳糖苷酶相关，因此其机制需要进一步研究。

三、其他酶的影响

研究表明，果胶和半纤维素中有很多半乳糖的存在，在细胞壁结构降解过程中，β-gal 会使带有支链的多聚醛酸发生降解，半乳糖含量降低，果胶被溶解。因此 β-gal 对果实软化也起到了一定的作用。Debra et al.（1992）研究了桃果实采后的软化过程，结果表明，β-gal 对果胶分子上半乳糖支链的水解会加速桃果实的软化。这在 Fan et al.（2019）对杏果实软化的研究中被证实。在番茄果实中沉默 β-gal 基因 *SITBG4* 可以有效抑制番茄果实软化（Smith et al., 2002）。陆玲鸿等（2022）研究了不同贮藏温度下猕猴桃果实软化相关酶活性的变化，贮藏温度为 25 ℃ 的猕猴桃果实软化与 β-gal 显著相关。

木葡聚糖作为一种半纤维素多糖广泛存在于细胞壁中。而木葡聚糖内转糖苷酶可以使木葡聚糖链发生断裂，细胞壁结构被降解，果实发生软化。有研究表明木葡聚糖内转糖苷酶对采后果实中半纤维素的降解起到主要作用。这一观点在蓝莓（Chen et al., 2017）、柿子（Wang et al., 2020）、梨（Chen et al., 2017）等果实软化研究中被证实。Lin et al.（2019）研究了壳聚糖处理对龙眼果肉的影响，结果表明，该处理可以有效降低木葡聚糖内转糖苷酶以及 β-gal 的活性，显著抑制龙眼果实的软化。

在果实成熟衰老过程中，细胞壁多糖的合成与交联作用使得细胞壁拥有良好的支撑作用，而细胞壁降解也导致了果实发生软化现象，因此对果实软化过程中细胞壁代谢进行研究，能显著地降低果实软化的速度，延长果实的货架期，对保持果实品质有很大的意义。

第六节　碳水化合物代谢

淀粉与糖作为果实细胞内最重要的两种内含物，其含量及成分对果实风味和采后果实代谢有很大的影响。淀粉降解对提高果实采后品质起着关键作用，这导致果实软化和变甜，并决定了消费者的接受度。有研究表明果实在成熟期间有大量淀粉累积，而这些淀粉会被淀粉酶降解为葡萄糖，生成的葡萄糖又会在异构酶的作用下变成果糖，果糖与葡萄糖在蔗糖磷酸酶的作用下合成蔗糖，蔗糖被转化酶分解为葡萄糖和果糖，最终淀粉与糖之间形成一种动态平衡，以达到维持果实硬度的目的（张强等，2020）。α-淀粉酶、淀粉分支酶、蔗糖磷酸合酶（Sucrose Phosphate Synthase, SPS）和 SS 是参与蔗糖合成的关键酶，能使采后猕猴桃的淀粉转化为糖（Liao et al., 2021）。蔗糖通过 NI 和酸性转化酶（Acid Invertase，AI）不可逆催化水解为葡萄糖和果糖（Lu et al.，2019）。

淀粉是果实主要的贮藏物质，也是细胞壁的支撑骨架，随着淀粉酶活性的提高，淀粉被降解为可溶性的葡萄糖和果糖，细胞的扩张力会降低，果实发生软化现象（艾沙江·买买提等，2018）。因此果实软化与淀粉的降解密切相关，这在苹果（齐秀东等，2015）、番荔枝（李伟明等，2018）、猕猴桃（陈金印等，2003）、香蕉（苗红霞等，2013）等果实软化研究中被证明。Mo et al.（2008）研究了番荔枝果实的成熟软化过程，结果表明，随着果实的不断成熟，果实硬度逐渐降低，这是由于淀粉被降解为可溶性糖。魏宝东等（2014）对磨盘柿果实软化进行了研究，结果表明，淀粉在淀粉酶的作用下被分解，所生成的糖为果实的呼吸跃变提供能量。当猕猴桃达到一定成熟度时，大量的圆形淀粉颗粒在细胞中积累，然后在随后的软化过程中逐渐降解（Nardozza et al.，2013）。胡丽松等（2017）对菠萝蜜果实的糖代谢过程进行研究，结果表明，随着淀粉酶活性的增强，大量淀粉被分解，果实发生软化。张强等（2020）在甜瓜果实后熟软化的研究中也得到了相同的观点，果实软化过程中淀粉酶活性显著提高，淀粉含量降低，果实硬度下降，因此淀粉水解是果实软化的重要原因。

第七节　激素

植物激素作为果实发育成熟的关键物质，通过各种复杂多样的生理生化过程影响果实的生长、成熟和软化。目前乙烯、脱落酸、生长素等植物激素对采后果实软化的调控作用在多种水果中均被证实。因此研究植物激素在果实软化中的调控作用显得尤为重要。

一、乙烯

乙烯是一种能调控植物生长发育的植物激素，长期以来，乙烯也是人们所公认的能调控果实成熟软化的激素。乙烯虽然在果实体内生成量非常微小但在植物生长发育等各个方面都起着重要的调节作用，能够诱导与植物组织衰老和果实后熟相关的一系列不可逆的生理生化进程。乙烯对跃变型果实和非跃变型果实的代谢活动影响基本一致，如促进呼吸，促进叶绿素、淀粉等物质的水解，促进胡萝卜素和花青素的合成等。

跃变型与非跃变型果实的重要区别在于其乙烯生成的特性和其对乙烯的反

应。跃变型果实后熟过程中一个明显的特征就是会大量生成乙烯，并伴随着呼吸高峰的出现，细胞膜透性的增加及与果实后熟相关的酶活性的不断增强，代谢物质转化急剧，果实日趋后熟和衰老；而非跃变型果实在后熟中却没有呼吸高峰的出现。根据果实对乙烯处理的反应，提出跃变型果实中乙烯生成有2个调节系统：系统Ⅰ负责跃变前果实中低速率乙烯的生成；系统Ⅱ负责调节伴随成熟过程中乙烯的自我催化和大量生成。非跃变型果实只有系统Ⅰ，没有系统Ⅱ。对于跃变型果实外源乙烯能启动系统Ⅱ，形成乙烯的自我催化，并且与所用的乙烯浓度关系不大，是不可逆反应；非跃变型果实则相反，外源乙烯在整个后熟期间都起作用，促进呼吸增加，其反应大小与所用的乙烯浓度相关，是可逆的，当外源乙烯去除后，呼吸即恢复到原有水平，同时不会促进乙烯增加。可见，人为地促进或抑制采后果实的内源乙烯生成，可加速或延缓果实的后熟软化进程。但也有研究认为，在有些果实的后熟软化过程中，乙烯只是一个决定果实后熟软化速度的因子，而非软化启动因子，这在桃（吴敏等，2003）和猕猴桃（陈昆松等，1999）等水果上均有报道。

有大量研究表明，乙烯可以有效促进呼吸跃变型果实的软化，而软化过程受乙烯合成、与相关受体结合启动信号转导以及下游基因来调控。乙烯合成途径是在1-氨基环丙烷-1-羧酸合成酶（Aminocyclopropane-1-carboxylate synthase，ACS）作用下，将S-腺苷甲硫氨酸（S-adenosyl-methionine，SAM）催化生成1-氨基环丙烷-1-羧酸，然后1-氨基环丙烷-1-羧酸在1-氨基环丙烷-1-羧酸氧化酶（1-aminocyclopropane-1-carboxylate oxidase，ACO）催化作用下合成乙烯（Lin et al.，2009），因此ACS和ACO是乙烯合成途径中的关键酶。有研究表明，*LeACS*2和*LeACO*1不仅参与了乙烯合成过程，当抑制其基因表达时还可以明显延缓番茄果实成熟（Wilkinson et al.，1995）。Ayub et al.（1996）也表明，反义ACO基因转入甜瓜可以提高甜瓜果实的硬度，有效延长甜瓜的货架期。Atkinson et al.（2011）发现，将猕猴桃果实用乙烯处理后，PG和果胶裂解酶基因表达显著增加，加快了猕猴桃果实软化。同样，对草莓果实进行外源乙烯处理后，*FaPG*1基因表达也显著提高，草莓硬度迅速降低（Villarreal et al.，2010）。因此乙烯可以通过促进细胞壁的降解来调控果实软化。Zhang et al.（2011）对牛油果进行外源乙烯抑制剂处理，研究发现，PG基因表达降低，显著抑制果实软化。Fan et al.（2018）对杏果实进行外源乙烯抑制剂处理，研究发现，乙烯受体抑制剂可以抑制细胞壁降解相关基因的表达，从而延缓果实软化。王慧等（2018）对柿果实进行纸片型1-MCP处理，结果表明，1-MCP处理可以有效降低柿果实呼吸强度，抑制柿果实采后软化。

ETH 受体作为乙烯信号转导途径中的负调控因子，通过其降解可以有效调控果实软化。Kevany et al.（2007）对番茄果实乙烯受体进行研究，发现 *LeETR4* 单突变体番茄对乙烯具有高敏感性，对番茄果实进行外源乙烯处理，可以使 LeETR4 快速降解，进而调控番茄果实成熟软化。*LeEIN2* 是乙烯信号转导途径中的正调控因子，Hu et al.（2010）通过沉默 *LeEIN2* 可以显著下调软化相关基因，达到延缓果实软化的目的。综上所述，乙烯对调控果实成熟软化有重要作用，可以通过抑制乙烯合成以及与受体结合来调控果实软化。

关于乙烯调控果实成熟衰老的作用机理，目前仍是研究的热点。不过有学者指出：①乙烯通过影响膜磷脂而改变了膜透性，使酶与底物的分区定位被打破导致呼吸反应加强；②乙烯能增加或控制细胞分泌或释放酶的速度，如能够增强 IAA 氧化酶，保护酶和纤维素酶的活力；③乙烯能促进果实成熟前蛋白质和 RNA 的合成并加速分解果实的衰老组织。影响乙烯合成的因素很多，如果实的成熟度、贮藏温度、贮藏气体条件等。一般来说，机械伤和病虫害会刺激乙烯的产生；适当的低温可以抑制果实乙烯的生成；适当地降低 O_2 浓度，提高 CO_2 浓度也会抑制果实乙烯的释放；另外，一些化学物质，如 1-MCP、AVG、AOA、Ag^+ 等会抑制果实乙烯的生成（彭丽桃等，2002）。

二、脱落酸

植物激素在果实发育成熟过程中起重要作用，呼吸跃变型果实的成熟软化依靠于乙烯的合成，而非呼吸跃变型果实的成熟软化依靠于脱落酸的合成，大量研究表明，脱落酸是调控非呼吸跃变型果实成熟软化的关键（李翠等，2023）。有研究表明，对桃果实进行外源脱落酸处理可以明显提高 PG 和 PME 的活性，使细胞壁结构发生降解，从而加快果实软化（郑秋萍等，2019）；纪迎琳等（2022）发现，外源脱落酸处理可以提高乙烯合成量，进而加快果实软化。Chen et al.（2016）对草莓果实进行外源脱落酸处理并对其进行转录组学分析，发现外源脱落酸处理后的草莓果实软化相关基因表达显著提高，并诱导了果实衰老相关基因的表达。Mattus et al.（2023）也发现，对草莓果实进行外源脱落酸处理可以显著提高 *FcPG*、*FcEXP5* 等细胞壁代谢基因的表达，加速果实软化。Jia et al.（2018）研究发现，对葡萄果实进行外源脱落酸处理既可以提高 PG 活性，促进果实软化，又可以提高花色苷含量，促进果实着色。此外，研究发现，外源脱落酸处理葡萄果实可以加快果实的软化，而对其进行脱落酸抑制剂处理可以有效推迟葡萄果实的成熟软化（杨方威等，2016）。因此探究脱落酸对果实软化的调控作用具有重要意义。

三、生长素

生长素在果实发育早期起着重要的作用，且果实的发育与生长素密切相关。目前植物生长素主要研究萘乙酸以及吲哚乙酸两类。有研究表明，生长素可以延缓果实的成熟，对草莓果实进行外源生长素处理，可以显著推迟草莓果实着色，并抑制 PG 基因的表达，延缓了草莓果实的软化（贾海锋等，2016）。Vendrell et al.（1969）对香蕉进行吲哚乙酸浸泡处理，结果表明，吲哚乙酸可以有效积累可溶性固形物含量，并延缓果实成熟。Purgatto et al.（2001）也证明了生长素通过抑制淀粉酶基因的表达，延缓果实软化。付润山等（2010）对柿果实进行萘乙酸处理，结果表明，萘乙酸处理可以有效降低细胞壁降解酶活性，达到延缓柿果实软化的目的。此外，生长素可以抑制编码 DNA 去甲基化酶的基因，从而维持果实的高甲基化水平，抑制果实成熟（Li et al.，2016）。研究发现，番茄果实中 *PpIAA*1 基因的过表达可以通过提高乙烯合成以及果实成熟软化相关基因的表达来加快果实成熟软化，缩短了番茄果实的货架期（Li et al.，2016）。在桃果实成熟过程中，生长素通过提高 *PpACS*1 的表达进而促进乙烯合成，使桃果实发生软化（Wang et al.，2021）。在猕猴桃果实中，调控生长素稳态基因 *AcGH*3.1 的沉默也可以提高果实硬度，延长果实货架期。对葡萄果实进行外源生长素处理，可以有效抑制脱落酸合成，降低软化相关基因的表达，从而维持果实硬度（Jia et al.，2017）。因此，生长素可以有效调控果实软化。

第三章 采后果实分子水平变化

果实的成熟软化过程非常复杂，包括一系列的生理生化过程，而该过程与大量基因的表达调控有关（Pesaresi et al.，2014），包括转录因子（TFs）。TFs 也称反式作用因子，是一种参与真核生物转录起始的蛋白，能够直接或间接识别并结合相互作用基因启动子上的特定顺式作用元件，从而在特定时间、空间和适宜强度下激活或者抑制靶基因的表达，进而影响植物生长发育过程（刘强等，2000；郭光艳等，2015；Guo et al.，2008；Liu et al.，1999）。TFs 广泛存在于所有真核生物中，在很多生物过程中发挥重要作用，如调节新陈代谢过程、调控细胞周期、控制生长发育、应答外界环境信号等。当植物受到非生物胁迫时，植物体经过一系列信号转导过程将感受到的刺激信号传入细胞核内，核内的 TFs 感知到此信号并调控下游抗逆相关基因的转录，在此过程中，TFs 有信号转换和放大的作用（马海珍，2018）。

TFs 的结构一般包括 DNA 结合域（DNA binding domain）、转录调控域（Transcription Regulation Domain，激活或抑制域）、核定位信号（Nuclear Localization Signal）和寡聚化位点（Oligomerization Site）4 个功能域（Johnston et al.，2015；Liu et al.，1999）。TFs 通过这些结构域与下游靶基因启动子上的顺式作用元件结合或通过与其他转录因子互作形成转录复合体来调控目标基因的转录，其中，DNA 结合域是同类型 TFs 中保守的一段氨基酸序列，能特异性识别并结合目标基因启动子序列上的结合元件。转录调控域决定了 TFs 正调控还是负调控靶基因。核定位信号是控制 TFs 入核的富含赖氨酸和精氨酸的一段区域（黄泽军等，2002；马海珍，2018）。寡聚化位点是一段保守的氨基酸序列，通常与 DNA 结合域形成一定的空间构象（马海珍，2018）。

越来越多的植物转录因子研究都有了较大的突破，因此果实成熟软化过程中转录因子的调控研究已成为新的热点之一（徐倩等，2014）。近年来，随着大量学者的深入研究，逐渐从果实中分析出多种参与采后果蔬成熟软化的转录因子，如 MADS-Box、MYB、AP2/ERF、NAC、WRKY、BZR、bZIP 等

(Peng et al., 2022；杨颖等，2009；范中奇等，2015），探究这些转录因子的调控机制对延缓果实成熟软化有着重要的意义。

第一节　AP2/ERF 转录因子对采后果实品质的调控

AP2/ERF 作为植物中最大的转录因子家族之一，含有 70 个氨基酸组成的 AP2/ERF 结构域（Nakano et al., 2006）。根据结构域的不同 AP2/ERF 可以分为 AP2、RAV、ERF、soloist 这 4 个亚类（Licausi et al., 2013），而 ERF 亚家族主要包括 DREB 和 ERF 这两类，其中 ERF 在果实成熟软化中的研究一直令人关注。ERF 转录因子参与乙烯信号转导途径，ERFs 通过与启动子区域上的顺式作用元件结合来调控乙烯响应基因的表达（Li et al., 2016；Liu et al., 2018）。起初在番茄果实中发现 LeERF1-LeERF4、LeERF3b 这几个 ERF 家族成员，且反义表达 LeERF1 能延缓番茄果实的软化，延长果实的贮藏期（Li et al., 2007）；而 LeERF2 在果实成熟软化过程中表达水平不断升高（Tournier et al., 2003），并且与 LeACO3 启动子上的顺式作用元件结合反馈调控乙烯的生成（Zhang et al., 2009）。ERF 转录因子参与了果实成熟软化，这在猕猴桃、番木瓜、香蕉、苹果等果实软化研究中均被证明（Xiao et al., 2013）。Yin et al.（2010）研究发现，猕猴桃的 ERF 转录因子能直接结合并激活果实软化相关基因的启动子，进而调控果实的成熟软化。猕猴桃 AcERF1B 和 AcERF073 均能结合 *AcGH*3.1 启动子并增强了猕猴桃中 *AcGH*3.1 的表达，降低了游离 IAA 含量，并增加了 IAA-Asp 含量。此外 AcERF1B 和 AcERF073 蛋白存在互作，而这种互作更增强了它们与 *AcGH*3.1 启动子的结合，表明 AcERF1B 和 AcERF073 通过激活 AcGH3.1 转录来正向调节 IAA 降解，从而加速猕猴桃采后成熟（Gan et al., 2021）。AP2a 作为 AP2/ERF 家族成员，可以负调控乙烯的合成，在番茄果实中沉默 AP2a 可以显著提高乙烯合成，促进番茄果实软化（Chung et al., 2010）。Wang et al.（2021）研究桃果实软化过程，研究发现，PpERF4 转录因子可以与 *PpACO*1、*PpIAA*1 基因启动子结合并激活其转录，转录后的 PpIAA1 与 PpERF4 相互作用形成复合物并激活果实软化相关基因的表达，进而调控桃果实软化。

第二节　EIN3/EIL 转录因子对采后果实品质的调控

EIN3/EIL 转录因子的氨基酸序列 N 端高度保守，包括酸性氨基酸区、碱性氨基酸区以及脯氨酸富集区等特征结构域。EIN3/EIL 转录因子能与启动子的 PERE 顺式作用元件相结合，激活并调控乙烯响应基因和细胞壁降解基因表达（Yang et al.，2015）。Yin et al.（2010）研究了猕猴桃果实的成熟衰老过程，结果表明，AdEIL2 和 AdEIL3 能结合细胞壁降解基因 *AdXET*5 和乙烯合成基因 *AdACO*1 的启动子并激活其表达，因此推测 AdEIL2 和 AdEIL3 能调控猕猴桃果实软化。有研究表明，甜瓜 CmEIL1 和 CmEIL2 转录因子可以结合 *CmACO*1 启动子并激活其表达，起到促进果实成熟软化的作用；苹果 MdEIL2 转录因子能与 *MdPG*1 启动子结合并激活其表达（Tacken et al.，2010）。

第三节 MADS-box 转录因子对采后果实品质的调控

酿酒酵母（Saccharomyces Cerevisiae）微小染色体维持蛋白 1（MCM1）、拟南芥（Arabidopsis Thaliana）AGAMOUS（AG）蛋白、金鱼草（Antirrhinum Majus）DEFICIENS（DEF）蛋白、现代人类（Homo Sapiens）血清应答因子（SRF）中均含有一个由 56~58 个氨基酸组成的高度保守的结构域，被命名为 MADS-box 结构域（王力娜等，2010；黄方等，2012）。包含该结构域的蛋白则被命名为 MADS-box 蛋白。MADS-box 结构域通常位于 MADS-box 转录因子的 N 末端，是 MADS-box 转录因子的 DNA 结合单元，负责识别和结合靶基因启动子区域的 CArG [CC（A/T）6GG] 盒。通常 MADS-box 蛋白间可通过相互作用结合成同源二聚体，有时也能与其他蛋白或辅助因子结合形成异源二聚体（刘菊华等，2010；黄方等，2012）。MADS-box 基因构成一个高度保守的转录因子家族，参与果实成熟软化的调控（Li et al.，2019）。MADS-box 转录因子含有 60 个氨基酸组成的 MADS-box 结构域，其主要分为 TypeⅠ型和 TypeⅡ型两个家族（Alvarezbuylla et al.，2000），TypeⅠ型家族分为 Mα、Mβ、Mγ、Mδ 4 个亚族，TypeⅡ家族编码的蛋白则具有典型的 MIKC 结构域，分为 MIKCc 和 MIKC* 两个亚族，作为调控果实成熟的关键转录因子被广泛研究。SEP 作为 MADS-box 亚家族成员，对调控果实成熟软化有重要作用。MADS 转录因子参与花器官发生、开花时间、胚胎和果实发育以及成熟衰老等一系列植物生长发育过程（Theißen et al.，2016）。Vrebalov et al.（2002）研究发现，成熟抑制因子 SiMADS-RIN 在番茄果实成熟软化的调控中起重要作用。在桃果实中 PrpMADS7 的沉默也显著延缓果实的成熟软化（Li et al.，2017）。Ito et al.（2008）研究发现，LeMADS-RIN 可以和乙烯合成基因 *LeACS*2 的启动子结合并激活其表达，控制乙烯的合成，进而影响果实软化进程。LeMADS-RIN

不仅能调控乙烯的合成，还能与细胞壁降解酶基因等其他下游基因发生反应（Qin et al.，2012），目前这一研究还有待深入研究。Qi et al.（2020）对甜樱桃（*Prunus avium* L.）果实成熟软化进行了研究，表明 PaMADS7 的差异表达变化与果实成熟进程一致，PaMADS7 通过与 *PaPG*1 启动子直接结合，正向调控 *PaPG*1 表达，促进甜樱桃果实成熟软化，且 PaMADS7 沉默能显著抑制甜樱桃果实的成熟，具体表现为果实硬度有所增加。综合以上研究结果表明 PaMADS7 在甜樱桃果实成熟软化的调控中发挥着不可或缺的作用。在香蕉果实软化研究中，通过沉默 MaMADS1 或 MaMADS2 可以有效抑制乙烯合成，从而达到抑制果实软化的目的（Elitzur et al.，2016）。利用 TRV 介导的病毒诱导的基因沉默技术沉默 *PaMADS*7 基因，抑制了果实成熟，影响了 ABA 含量、可溶性糖含量、果实硬度、花青素含量等主要成熟相关生理过程以及成熟相关基因的表达；酵母单杂交和瞬时表达实验证实甜樱桃 PaMADS7 能直接结合 *PaPG*1 的启动子并激活其表达，表明 PaMADS7 是甜樱桃果实成熟软化不可缺少的正调控因子（Qi et al.，2020）。SEPALLATA（SEP）基因是 MADS-box 亚家族，对调控花器官发育、开花时间和果实发育成熟具有重要作用。桃 PrupeSEP1 编码的氨基酸序列与苹果 MdMADS8 和 MdMADS9 以及草莓 MADS-RIN-like 非常相似，在桃贮藏期间，SEP1 的表达模式与乙烯生物合成和乙烯信号转导相关基因的表达模式相似（EIN2 和 ETR2），SEP1 的表达水平与 EIN2 和 ETR2 的表达水平具有相关性。此外，SEP1 的表达模式与细胞壁修饰相关基因（PG3、EXP2 和 PME3）相似，通过病毒诱导的转基因沉默技术（VIGS）抑制 SEP1 表达后，与对照相比，果实保持坚硬，果实软化延迟，而成熟软化相关基因 ACS2、EIN2、PME1、PG3、ACO1、和 Lox1 的表达水平显著降低。通过酵母单杂交实验发现 SEP1 能直接与 PG2 和 PG3 的启动子结合，促进其在果实成熟软化过程中的表达（Li et al.，2017）。苹果 SEP1/2-like MdMADS8 转录因子能够激活参与苹果成熟软化的 PG1 的转录，介导果实软化（Ireland et al.，2013）。抑制番茄 MADS-box 转录因子 SlBMP8 表达后发现，细胞壁代谢相关基因 PG，EXP，HEX，TBG4，XTH5 和 XYL 被诱导表达，果实失水更快，耐贮藏性降低，酵母双杂交表明 SlBMP8 与 SlMADS-RIN 互作（Yin et al.，2017）。在 SlMBP3 基因敲除的果实生长发育过程中，细胞增大相关基因（β-扩展蛋白基因 SlEXPB1 和内切 β-1,4-D-葡聚糖酶基因 Cel8）和果胶酶抑制剂相关基因（果胶酯酶抑制剂基因 PE 抑制剂和 PG 抑制剂基因 PG 抑制剂）表达上调，果胶酶编码基因（PG 基因 QRT3-like 和果胶裂解酶基因 PL2）表达下调（Kim et al.，2022）。

第四节 MYB 转录因子对采后果实品质的调控

高等植物含有大量的 TFs，其中 MYB 是最大的 TFs 家族之一，约占总转录因子的 9%，并参与植物生长、发育、代谢和胁迫响应等许多方面的调控（Xing et al.，2019；Pabo et al.，1992；Riechmann et al.，2000）。v-Myb 是第一个被分离出的 MYB TFs，来自于一种动物病毒，后经序列分析发现，该 TFs 可能起源于脊椎动物，突变之后成为了病毒的基因（Klempnauer et al.，1982）。不仅在脊椎动物中发现了 V-Myb 基因，也在植物、昆虫和真菌中发现了 A-MYB、B-MYB 和 c-MYB 等与其相关的基因，这些基因在细胞增殖和分化过程中具有重要作用（Weston，1998；Lipsick，1996）。第一个在植物中被鉴定出来的 MYB TFs 是一种在合成花青素过程中起重要作用的玉米 MYB TFs，编码一种含有 MYB 结构域的蛋白（Pazares et al.，1987）。MYB TFs 都具有保守的 DNA 结合域，即 MYB 结构域。MYB 结构域由几个重复序列组成，每个重复序列有 50~53 个氨基酸残基，通过分析 MYB 域的三维结构，可以发现每个重复序列的第二个和第三个螺旋具有三个规则间隔的色氨酸（W）残基，形成带有疏水核心的螺旋-转-螺旋（HTH）结构，而第三个螺旋则直接与目标 DNA 的主沟相互作用（Ogata et al.，1996；Jia et al.，2004）。根据结构域重复序列数目，MYB 超家族分为四个主要亚家族，包括 MYB（1R-MYB，1个重复）、R2R3-MYB（2R-MYB，2 个重复）、R1R2R3-MYB（3R-MYB，3 个重复）和 4R-MYB（4R-MYB，4 个重复），其他 MYB 的分类通过与 R1，R2 和 R3 比较相似性来确定，在这些类型的 MYB TFs 中，R2R3-MYB 是主要的亚家族（Stracke et al.，2001）。

MYB 转录因子在果实发育和成熟软化过程中发挥重要作用。在番茄和草莓果实中均鉴定出 MYB 转录因子，其中一些转录因子在细胞壁代谢以及次生代谢中起着重要作用（钱丽丽等，2023）。Cao et al.（2020）发现，番茄果实中的转录抑制因子 SiMYB70 通过与 *SlACS2* 基因启动子结合，抑制其转录，达到延缓番茄果实软化的目的。Liu et al.（2021）发现，番茄果实中 *SiMYB75* 过表达可以显著下调 *SlFSR* 表达，有效延长果实货架期。Cai et al.（2021）研究发现，过表达 FvMYB79 可以使 *FvPME38* 表达显著上调，导致草莓果实软化加快，当沉默 *FvMYB79* 时，草莓果实硬度明显升高。芒果 MiMYB8 通过直接结合苯丙氨酸解氨酶（Phenylalanine ammonia lyase，PAL）*MiPAL1* 启动子来抑制其转录，在烟草叶片和芒果果实中瞬时过表达 MiMYB8 发现其可通过降低

PAL 活性和下调该基因表达来抑制花青素的积累，表明 MiMYB8 可能在芒果果实成熟过程中通过负调节 *MiPAL* 基因起到抑制花青素合成的作用（Aslam et al.，2024）。MaMYB3 可以抑制香蕉中的淀粉降解酶基因的表达，进而调控香蕉果实的成熟软化（Fan et al.，2018）。此外，CpMYB1 和 CpMYB2 还能结合木瓜细胞壁降解酶基因 *CpPME*1、*CpPME*2 和 *CpPG*5 的启动子，通过调控这些基因的表达，参与木瓜果实软化（Fu et al.，2020）；进一步研究发现，CpMYB1 和 CpMYB2 均为转录抑制子，能抑制 *CpPME*1、*CpPME*2、*CpPG*5 启动子的活性，这一发现为研究 MYB 转录因子在果实软化中的作用提供了新的思路。

第五节　NAC 转录因子对采后果实品质的调控

　　NAC（NAM，ATAF1/2，CUC2）是植物特有的一类 TFs（Duval et al.，2002），也是植物 TFs 家族中成员最多的家族之一。NAC TFs 的 DNA 结合域由 150 个保守的氨基酸组成，该结合域可以分为 A、B、C、D、E 5 个结构域，A、C、D 结构域高度保守，而 B、E 保守性不强，因此 NAC TFs 具有功能多样性（Olsen et al.，2005）。

　　NAC 转录因子在一系列生理过程如叶片衰老、花形态发生、果实发育、成熟、衰老、次生代谢、应激反应、生物和非生物胁迫的防御反应中都发挥重要作用（Liu et al.，2022；Olsen et al.，2005；Marques et al.，2017）。番茄突变体的发现揭示了 NAC 转录因子能抑制乙烯产生、果实软化和色素积累，从而延迟果实成熟（Mizrahi et al.，1976；Yuan et al.，2016；Kumar et al.，2018）。研究发现，PavNAC56 可直接与细胞壁代谢相关的多个基因（*PavPG*2、*PavEXPA*4、*PavPL*18 和 *PavCEL*8）的启动子结合，并激活它们的表达，表明 PavNAC56 在控制甜樱桃果实的成熟和软化过程中起着不可或缺的作用。柿子 DkNAC9 已被鉴定为 *DkEGase*1 启动子的反式激活子，能加速果实软化（Wu et al.，2020）。PavNAC56 还调节与花青素生物合成和其他果实成熟相关过程相关基因的转录水平（如 *PavPAL*、*PavCHS*、*PavANS*、*PavUFGT*、*PavGG*1、*PavPG*1 和 *PavWiv-*1），最终影响果实成熟过程中的颜色和硬度（Qi et al.，2022）。NOR（non-ripening）是一种 NAC 转录因子，在苹果中，NOR 同源物 MdNAC18.1 被认为是果实软化和决定采收时期的主要因素（Migicovsky et al.，2021）。在草莓中，NAC 转录因子 FaRIF/FaNAC035 通过调节细胞壁降解、果实软化、色素和糖积累等过程，从而控制果实成熟（Martín-Pizarro et al.，

2021）。FcNAC1 通过影响果胶代谢控制桃果实的软化（Carrasco-Orellana et al.，2018）。香蕉 NAC 转录因子 MaNAC029 可转录激活与乙烯生物合成和与果实品质形成有关的多种细胞代谢相关的基因表达（细胞壁降解、淀粉降解、香气化合物合成、叶绿素分解代谢）调节果实成熟，在香蕉中过表达 MaNAC029 可激活乙烯生物合成，加速果实成熟和品质形成，当 E3 连接酶 MaXB3 与 MaNAC029 蛋白相互作用后可促进 MaNAC029 蛋白酶体降解，表明 MaXB3 过表达会减弱 MaNAC029 对乙烯生物合成和品质形成的影响（Wei et al.，2022）。猕猴桃 AcNAC1 和 AcNAC2 通过激活木葡聚糖内转葡萄糖化酶/水解酶 *AcXTH*1 和 *AcXTH*2 基因来调控果实软化（Fu et al.，2024）。桃 PpNAC1 和 PpNAC5 协同激活果胶酶 *PpPGF* 的转录调控果实成熟过程中的软化（Zhang et al.，2024）。以上研究均证实 NAC 转录因子在采后果实品质方面有重要的调控作用。

第六节　其他转录因子对采后果实品质的调控

WRKY TFs 只在植物中存在，因含有保守的 WRKYGQK 序列因此被称为 WRKY TFs，其 C 端还存在锌指结构域（Yamasaki et al.，2005）。根据 WRKY TFs 所含 WRKY 结构域数量和 C 端锌指结构特征将其分为 3 个亚家族，含有两个 WRKY 结构域和 C2H2 锌指结构的为第Ⅰ类，含有一个 WRKY 结构域和 C2H2 锌指结构的为第Ⅱ类，含有 C2-HC 锌指结构的为第Ⅲ类。WRKY TFs 能特异性识别并结合 W-box（TTGAC），从而调控下游靶基因的转录。WRKY 在抵御生物（Kalde et al.，2003）和非生物胁迫（Mare et al.，2004）中起重要作用。将 LcWRKY47 基因转入完整的荔枝果实后发现，转基因果实中 LcWRKY47 的表达量显著高于对照果实，贮藏第 3 d 时果实几乎没有褐变的发生，而对照出现了严重的褐变现象，表明 LcWRKY47 转录因子可能通过延缓荔枝的衰老与褐变来维持果实采后品质（程南谱等，2024）。果实成熟过程中淀粉的降解与苹果的老化过程和风味的形成密切相关，*MdBams* 在苹果淀粉-糖转化过程中发挥着重要作用，瞬时过表达 *MdBam5* 显著降低了苹果中的淀粉含量，此外 MdWRKY32 可直接结合 *MdBam5* 启动子并激活 *MdBam5* 的表达，参与调控苹果采后淀粉-糖代谢（Li et al.，2021）。研究发现，猕猴桃 AcWRKY40 启动子中有多个与成熟和衰老相关的顺式作用元件，GUS 活性分析表明 AcWRKY40 启动子活性受外源乙烯诱导，酵母单杂交和双荧光素酶实验表明 AcWRKY40 可以与 *AcSAM2*、*AcACS*1 和 *AcACS*2 启动子结合并激活它们，瞬时转化实验表明AcWRKY40可以增强 *AcSAM2*、*AcACS*1 和 *AcACS*2 的表达，表明AcWRKY40可能通过调控乙烯生物合成相关基因的表达，参与猕猴桃采后成熟过程（Gan et al.，2021）。猕猴桃MaWRKY49可通过正调控果实成熟过程中果胶酸裂解酶 *MaPL*3 和 *MaPL*11 基因表达参与果实的成熟与软化（Liu et al.，2023）。

bZIP TFs是广泛存在于植物中的一类TFs，在不同物种中根据保守结构域可以将bZIP TFs分为不同的亚家族。bZIP TFs能特异性结合下游目标基因启动子上的ACGT核心元件，如G-box（CACGTG）、C-box（GACGTC）和A-box（TACGTA）（Izawa et al.，1993），这些TFs在植物种子萌发、光信号应答、以及各种生物和非生物胁迫中起重要作用（Shrestha et al.，2014）。全基因组分析显示，LcbZIP7、LcbZIP21、LcbZIP28、LcbZIP1、LcbZIP4可能参与调控采后贮藏期间荔枝果实的衰老，LcbZIP40/41可能在荔枝果实对病原体感染的反应中发挥重要作用（Hou et al.，2022）。柿子bZIP转录因子DkTGA1能结合脱涩相关基因 *DkADH*1、*DkPDC*2 和 *DkPDC*3 的启动子，但激活能力一般，DkERF9与DkTGA1组合后对 *DkPDC*2 启动子的激活表现出附加效应，表明DkTGA1可通过间接结合 *DkPDC*2 启动子对采后柿子涩度品质起作用（Zhu et al.，2016）。

bHLH转录因子超家族蛋白质是由一个高度保守的bHLH结构域定义的，该结构域包含两个不同的功能区：Basic区和HLH区（Li et al.，2006）。Basic区位于bHLH结构域的N端，是一个由大约15个氨基酸组成的DNA结合区，该结合区一般能够结合到特定的E-box（CANNTG），进而调控靶基因的表达（Atchley et al.，1997）。酵母单杂交、电泳迁移率分析和双荧光素酶报告基因分析表明，番木瓜CpbHLH3可直接结合木聚糖酶CpXYN1启动子并激活其转录，将CpbHLH3瞬时过表达后可使木聚糖酶活性增加，表明CpbHLH3可通过调节CpXYN1的表达参与采后番木瓜果实的软化（Wang et al.，2022）。

在猕猴桃中，锌指AdZAT5正向调控果胶降解相关基因 *Adβ-Gal*5 和 *AdPL*5 的启动子（Zhang et al.，2022）。苹果锌指MdZF-HD11通过促进 *Mdβ-GAL*18 的表达来调节果实软化（Wang et al.，2023）。BZR（BRASSINAZOLE-RESISTANT）家族基因是编码参与油菜素内酯信号转导的植物特异性转录因子。在香蕉果实中，MaBZR1/2与MaMPK14相互作用，促进果实成熟过程中细胞壁基因如 *MaXET*5、*MaEXP*2 和 *MaPL*2 的转录抑制（Shan et al.，2020）。以上研究结果表明转录因子可通过与靶基因结合从而调控采后果实的品质变化，对采后果实的生理生化变化起重要调控作用。

第二篇

采后果实的调控研究

第四章　物理保鲜处理对采后果实品质的影响

第一节　低温处理对采后果实品质的影响

许多果蔬采摘后，由于呼吸代谢旺盛，常温下贮藏 2~3 d 即开始发生黄化、软化、腐烂变质，严重降低其商品品质（刘学等，2015）。温度影响果实新陈代谢，所以是保持品质、延长货架期的关键因素。低温贮藏是现今在果蔬采后保鲜技术中最常用的方法之一（拓俊绒等，2005），是通过降低样品所处的环境温度从而抑制果实的呼吸作用，降低果实的能量损失，减缓果实内各种新陈代谢和微生物生长，减弱生命活动，降低果实营养成分的消耗，让果蔬始终处于一个较低温的贮藏环境中，延长贮藏期及降低腐烂率，可有效地维持果蔬采后外观和营养品质，从而让果实保持较长的保质期（赵颖颖等，2012；陈琪琪等，2023）。Sang et al.（2022）发现，低温贮藏能延缓冬枣果实硬度下降、推迟果皮转红，较好地维持其外观品质，显著降低呼吸速率、失重率和腐烂率，抑制可溶性固形物、可滴定酸和抗坏血酸的减少，从而维持其营养品质。Sang et al.（2021）研究发现，0 ℃条件下贮藏冬枣可以显著提高抗氧化酶活性和相关基因表达量，保持冬枣的采后品质和抗氧化能力。研究表明，甜玉米在低温（4 ℃）贮藏条件下呼吸强度和淀粉含量的下降被显著抑制，可溶性糖含量保持较高水平，较好地保持了其营养品质（Calvo-Brenes et al.，2020；李朝森等，2018；单秀峰等，2015）。儿菜在低温贮藏下可保持良好的外观品质，避免质量、硬度、硫代葡萄糖苷和可溶性糖含量下降（Sun et al.，2020）。与常温贮藏相比，2 ℃和 5 ℃低温贮藏能显著降低椰果的呼吸速率，有效降低失重率、外壳褐变、MDA 含量等，并将贮藏期从 3 d 延长至 6 周（Luengwilai et al.，2014）。Ge et al.（2017）发现在低温条件下，枇杷 *EjNAC*3 和 *EjNAC*4 被诱导表达，并能结合果实木质素合成相关基因 *EjCAD-like* 启动子激活其转录，程序降温和热水处理能抑制 *EjNAC*3 和 *EjNAC*4 的转录和枇杷果实木质素的含量。陈熙等（2024）比较了 0 ℃、4 ℃和 20 ℃贮藏温度下马铃薯品质及能量代谢之间的关系，发现与 20 ℃处理相比，4 ℃贮藏可抑制马铃薯茎肉硬度下降和呼吸强度上升，保持马铃薯块茎中能量代谢酶 H^+-ATPase、Ca^{2+}-ATPase、细胞色素氧化酶和 SDH 的活性，提高马铃薯块茎组织内 ATP、二磷酸腺苷（Adenosine Diphosphate，ADP）、磷酸腺苷（Adenosine Monophosphate，AMP）含量及能荷；与 4 ℃处理相比，0 ℃处理中马铃薯呼吸强度增强、可溶性总糖含量升高，并导致能量代谢相关酶活性下降，使能量代谢水平下降，说明 4 ℃贮藏温度能维持马铃薯块茎较高的能量水平，保持较好的贮藏品质。但一些冷敏型果蔬如桃（赵颖颖等，2012）、油桃（高慧等，2010）、橄榄（孔祥佳等，2011）、茄子（赵云峰等，2010）、黄瓜（袁蒙蒙等，2012）等在较长期的低温储存时极易发生冷害。表明低温贮藏可延缓一些果实采后品质，但某些果实不适用。

第二节 气调保鲜对采后果实品质的影响

气体环境的组成是影响果实保鲜的重要因素之一，气调贮藏是通过改变贮藏环境中的气体成分进而影响果实的贮藏品质。其中，对气体成分的调控主要是降低 O_2 浓度、增加 CO_2 浓度以及排除乙烯等，是一种高效的果蔬保鲜方法（Yahia et al.，2019）。气调贮藏包括被动气调（Controlled Atmosphere，CA）和自发气调（Modified Atmosphere，MA）2 种，其中 MA 是通过利用果实自身的呼吸代谢来调节周围的气体成分，从而达到低 O_2 高 CO_2 的环境，抑制果实呼吸代谢，减轻果实品质的下降。CA 是按照标准来严格人为控制其周边的气体成分（何伟，2020）。气调可以有效地抑制果实呼吸作用和微生物生长，延缓果实衰老和软化，并且减少乙烯产生（Beaudry，1999；Weber et al.，2019；Nakata et al.，2020）。通过调节 O_2 和 CO_2 的比例是气调贮藏中最为常见的一种气调方式（郭慧静等，2023）。目前，气调贮藏技术已被广泛应用于苹果、梨、猕猴桃和番茄等果蔬的贮藏保鲜上（杜艳民等，2021；Arshad et al.，2022）。鲁奇林等（2014）发现，5% O_2+2% CO_2+93% N_2 气调包装可显著抑制冬枣在贮藏期间维生素 C、cAMP 和总黄酮的流失，保持果实色泽，贮藏期可达 90 d。有研究表明，利用自发气调技术可以有效延缓'海沃德'猕猴桃在冷藏以及货架期间的软化（Han et al.，2022），同时还可以减缓冷藏过程中果实的重量损失（Ozturk et al.，2019）。荔枝经气调包装（5% CO_2+5%

O_2+90% N_2）处理后于 4 ℃±1 ℃条件下贮藏，可有效维持可溶性固形物和可滴定酸的含量，抑制果实自溶和果皮褐变，提高 SOD、CAT、APX、POD 的酶活性，表明该气调处理可维持荔枝贮藏品质，提高荔枝采后耐贮性（黄方等，2022）。降低氧气（O_2）和升高二氧化碳（CO_2）的自发气调包装处理可显著提高抗葡萄果实的氧化能力，从而保持果实的硬度并降低腐烂率，维持果实品质（Khalil et al.，2024）。5% O_2 体积分数 MA 气调贮藏方式可减轻湖景蜜露和中华寿桃果实冷害的发生（徐思朦等，2023）。5%~6% 的 O_2+0%~1% 的 CO_2 是黄金梨的理想贮藏条件，可有效减少乙醇、乙醛等物质的积累，延缓梨果实的衰老，但一旦环境中 O_2 体积分数低于 5% 时，则会诱发果实无氧呼吸并加速黄金梨的酒软（田龙，2007）。玉露香梨最佳气调参数为 1% O_2+3% CO_2（贾晓辉等，2023）。在蓝莓贮藏期间充入 7% O_2+20%CO_2+73% N_2 可明显提高果实的硬度和固酸比，降低失水率和烂果率，抑制乙烯生成速率，表明气调包装可有效改善蓝莓贮藏期间的果实品质（孙也等，2024）。李宁等（2016）通过对草莓进行气调包装（5% O_2+10% CO_2+85% N_2）并在 0 ℃条件贮藏，可以有效保持草莓果实的良好品质，保鲜期延长至 30 d 左右。自发气调包装袋维持 O_2 体积分数为 13%±3%，CO_2 体积分数为 8%±1%，相对湿度为 95%±1% 时，可有效降低杨梅营养物质损失和坏果率，减缓硬度下降，有效延长新鲜杨梅的货架期（郑鹏蕊等，2022）。以上研究结果表明，不论 CA 还是 MA，都对果蔬保鲜有一定的效果。

第五章　化学保鲜处理对采后果实品质的影响

化学保鲜是指使用化学保鲜剂来提高果蔬的耐贮性，从而延长果蔬保鲜期的技术。化学贮藏由于方法简单易行，效果稳定显著而十分受关注，它既可单独使用，又可作为冷库的辅助措施，是近年保鲜技术发展的一个重要方面（阿地拉·阿不都拉，2021）。

第一节　正丁醇对采后果实品质的影响

细胞膜是最初感知冷胁迫的界面结构。果蔬在冷害条件下，由于不饱和膜脂含量减少，膜透性增加，细胞内电解质和可溶性物质外渗，与膜结合的酶活性逐渐降低，酶促反应失调，导致细胞膜损伤，从而表现出冷害症状（Shan et al.，2022）。细胞膜中的脂质由三种主要成分组成，包括磷脂、糖脂和胆固醇。磷脂是最重要的组成成分，在信号传导、膜流动和细胞骨架组成中发挥关键作用。研究表明，正己醛、正丁醇和仲丁醇是 PLD 活性抑制剂，能够抑制磷脂酸以及磷脂降解，从而保持细胞膜的完整性，维持果蔬品质，已被应用于龙眼（Li et al.，2019）、桃（Wan et al.，2019）、草莓（李英华，2009）和桑葚（Bernardo et al.，2021）等采后水果的保鲜。其中，正丁醇是由 PLD 催化磷脂酰基产生的一种底物，是 PLD 的特异性抑制剂。正丁醇的存在可导致 PLD 在转磷酰化反应中反应更快，从而抑制了 PLD 的水解活性（Dhonukshe et al.，2003）。此外，也有研究表明 PLD 活性被正丁醇抑制的机理是正丁醇催化效果优于水，当正丁醇作为催化 PLD 的底物且其与 PC 反应时，能够生成水以及稳定的磷脂酰肌醇。正丁醇也可通过直接抑制 PA 从而抑制 PA 的下游反应和细胞膜膜脂过氧化，从而起到保护细胞膜的功能（Dek et al.，2018）。新疆哈密瓜（*Cucumis melo* 'Hami'）被列为新疆第一大水果，深受广大消费者的青睐。哈密瓜属于典型的呼吸跃变型果实，其采收期较集中，通常利用低温来延长其新鲜度时期，但哈密瓜对低温敏感，不适宜的低温易使果实发生冷害，冷害症状最初表现为表面出现深色或浅褐色斑点，随后组织表皮出现凹陷状斑点（Huang et al.，2023）。用正丁醇处理哈密瓜能降低果皮 MDA 含量，细胞膜透性，PLD 活性及 *CmPLD-β* 表达，LOX 活性及 *CmLOX* 表达；提升了膜脂组分 PI 含量和不饱和脂肪酸亚油酸、亚麻酸、油酸和芥酸含量，抑制了 PA 的产生和饱和脂肪酸硬脂酸的升高，缓解了果实冷害的发生（刘彩红，2022；Huang et al.，2023）。正丁醇处理桃果实抑制了膜脂过氧化产生的 MDA、细胞膜透性和 *PpPLDα* 的表达，增强了桃果实的抗寒性（Wan et al.，2019）。用正丁醇处理荔枝（*Litchi chinensis*）可以保持较高的抗坏血酸含量，并降低冷冻荔枝的冷害、霉菌活性、褐变和 PLD 的活性（Bhushan et al.，2019）。南果梨是辽宁省的特产水果，由于其色泽鲜艳、果香浓郁、肉质细腻、酸甜适口、营养丰富而深受消费者喜爱。采后南果梨在常温贮藏过程中会逐渐后熟衰老，果心逐渐出现褐变，严重影响了南果梨的商品品质和价值。果实褐变主要是由酶促褐变引起的，酶促褐变的三要素包括氧气、酶和底物。PPO 是酶促褐变的主要酶，酚类物质酶促褐变的主要底物，通常情况下 PPO 与酚类物质被膜分离，彼此不会接触到，当果蔬受到生物胁迫或非生物胁迫后，细胞膜脂代谢受到影响，发生膜脂过氧化和降解现象，细胞膜的结构被破坏，PPO 与酚类底物的区室化分布被破坏，二者发生接触，PPO 将酚类物质氧化为醌类物质，导致果实发生褐变（Ali et al.，2018）。研究发现正丁醇处理南果梨后能抑制果心组织 PLD 和 PPO 酶活性及 PLD 家族成员表达水平，进而抑制磷脂的水解以及亚油酸和亚麻酸两种不饱和脂肪酸的氧化进程，减少膜脂过氧化产物 MDA 含量，降低膜脂过氧化程度，保持了细胞膜和细胞器结构的完整性，降低了常温贮藏过程中果实褐变指数和褐变率，有效延缓了褐变的发生；正丁醇处理也延缓了南果梨果实硬度的下降和可溶性固形物含量的增加，缓解了贮藏过程中果实衰老的进程，较好地维持了果实品质（孙扬扬，2020）。以上研究表明正丁醇作为 PLD 抑制剂可通过维持采后果实膜脂代谢对其起到保鲜的作用。

第二节　1-MCP 处理对采后果实品质的影响

1-MCP 是一种无毒、无味、无色气体，1-MCP 能够与乙烯受体结合，从而阻止乙烯与受体结合，进而抑制乙烯诱导的一系列生理生化反应，如果实成熟和衰老过程（冯叙桥等，2013）。软枣猕猴桃是一种很有吸引力的水果，因其体积小、皮肤光滑、味道独特而越来越受到消费者的关注（Wang et al., 2015；Xu et al., 2021），果实中富含抗坏血酸（AsA）、叶黄素、总酚类物质（TP）和矿物质，特别是磷、钙、铁和锌（Krupa et al., 2011）。它还具有营养特性，如抗炎、抗氧化、抗癌和降低血压（Latocha et al., 2010）。采摘后的软枣猕猴桃在20℃放置5 d就会软化（Sutherland et al., 2021），这主要由于其属于呼吸跃变型果实，会产生乙烯成熟激素，当到达呼吸跃变时，呼吸、成熟和软化也会加速（Wang et al., 2015）。果实软化的特征是细胞壁结构中的一系列多糖组分如果胶、纤维素和半纤维素瓦解（Fan et al., 2019）。果胶是细胞壁中的重要组分，对维持细胞结构的完整性和增强细胞间粘附性方面其重要作用。利用不同的提取试剂，果胶可以分为水溶性果胶（Water-soluble Pectin，WSP）、CDTA溶性果胶（CDTA（Trans-1, 2-cyclohexanediamine Tetraacetic acid）-Soluble Pectin，CSP）、碳酸钠溶性果胶（Sodium Carbonate-Soluble Pectin，SSP）（Holland et al., 2012）。在果实成熟的过程中纤维素和半纤维素也会发生改变，纤维素、果胶和半纤维素结合形成网状结构，作为细胞壁的组成部分（Bu et al., 2013）。硬度的降低是果实软化的主要表现之一，是由各种生理代谢活动的变化、水分流失、细胞壁结构解体、果胶降解和膜损伤引起的（Ji et al., 2021）。1-MCP处理的软枣猕猴桃果实硬度、可滴定酸度、抗坏血酸、总酚类物质和黄酮类物质含量均保持较高的水平，用电子鼻和电子舌测定发现1-MCP处理后长期冷藏的水果有更高的感官得分。值得注意的是，1-MCP通过降低细胞壁修饰酶的活性，延缓了细胞壁成分如果胶、纤维素和半纤维素的降解。此外，1-MCP降低了碳水化合物代谢相关酶的活性，导致果实中淀粉和蔗糖含量水平较高，葡萄糖、果糖和山梨醇含量水平较低。1-MCP处理可通过抑制番木瓜果实木聚糖酶活性，提升果实半纤维素含量，从而延缓采后番木瓜果实软化的发生（Wang et al., 2022；Xiong et al., 2023）。综上所述，1-MCP可以延缓黄花蒿果实的软化，延长其贮藏期和保质期。

第三节　甜菜碱处理对采后果实品质的影响

甘氨酸甜菜碱（Glycine Betaine，GB），是一种季铵化合物，具有重要的渗透调节作用，在胁迫条件下维持植物细胞膜、蛋白质或酶的稳定性，从而保护组织免受非生物胁迫，增强抗逆性（Wang et al., 2016；Zhang et al., 2016）。在高等植物中，GB具有保护细胞膜和稳定蛋白质或酶的功能，从而增强对低温和盐胁迫的耐受性（Park et al., 2006）。研究发现，低温胁迫过程中植物内源甜菜碱的积累与耐逆性增强（Kishitani et al., 1994）密切相关。Xing et al.（2001）研究发现，拟南芥叶片中的内源甜菜碱含量在冷驯化过程中随着植株抗冻性的增加而急剧提高。近年来，关于外源 GB 处理对果蔬的影响已有一些报道，包括桃（Shan et al., 2016；Wang et al., 2019a；Wang et al., 2019b）、西葫芦（Yao et al., 2018）、山楂（Razavi et al., 2018）、枇杷（Zhang et al., 2016）、甜椒（Wang et al., 2016）和梨（Luo et al., 2020；Sun et al., 2020；Wang et al., 2020）、石榴（Molaei et al., 2021）、香蕉（Chen et al., 2021）。

陈丽娟等（2015）研究发现，8 mmol/L 甜菜碱处理均可较好保持香椿嫩芽品质，至贮藏结束，仍具有商品价值，且能有效减轻香椿嫩芽的质量损失和腐烂现象，减缓叶绿素、维生素 C 和总黄酮含量损失，降低呼吸强度，抑制多酚氧化酶（PPO）活性，从而延长其保存期限。冬枣果实在收获后很容易变软，将冬枣浸泡在 20 mmol/L 的 GB 中 20 min，发现 GB 处理可以通过抑制 *PG*、*CX*、*PME* 和 *β-Glu* 等基因的表达和酶活性，有效地维持细胞壁成分的含量。同时，GB 处理可提高抗氧化酶（APX、CAT、SOD、POD）活性和非酶抗氧化剂（MDA、H_2O_2、ASA、GSH）含量，进而降低了 ROS 含量。此外，能量代谢酶（H^+-ATPase、Ca^{2+}-ATPase、SDH 和 CCO）活性和基因表达均显著增加，从而维持了较高的能量水平（ATP、ADP、AMP 和 EC）说明 GB 可通过增加能量代谢来增强冬枣体内 ATP 的生物合成，为抗氧化代谢提供了必要的能量，延缓了采后枣的软化（Zhang et al., 2023）。有研究表明，GB 处理显著降低了枇杷果实的 MDA 含量，并通过诱导抗氧化酶活性提高了贮藏耐受性（Zhang et al., 2016）。GB 浸泡处理可以通过增加抗氧化酶的活性来减少石榴果实的冷损伤，以保持良好的营养质量（Molaei et al., 2021）。GB 还可以通过增加能量代谢相关酶的活性来提高番木瓜果实的耐寒性（Pan et al., 2019）。应用 GB 外源处理"摩洛"血橙果实，在 3 ℃贮藏 90 d 并随后在 20 ℃常温贮藏 2 d，对贮藏期间果实的生物活性、抗氧化活性和理化性质进行了评价，发现 GB 处理显著降低了冷藏期间"摩洛"血橙果实的质量和硬度损失，GB 处理还保持了较高浓度的有机酸（柠檬酸、苹果酸、琥珀酸和草酸）和糖（蔗糖、葡萄糖和果糖）的浓度，特别是对于较高剂量的 GB 处理（30 mmol/L）。在贮藏过程中，GB 处理还提高了血橙中总花青素浓度、总酚含量和总抗氧化活性。在酶活性方面，外源 GB 还提高了血橙中 CAT、APX、SOD、苯丙氨酸解氨酶含量升高，而多酚氧化酶活性受到抑制。总的来说，最有效的处理方法是 30 mmol/L GB，从而在长期储存期间维持"摩洛"血橙果实的生物活性化合物、抗氧化活性和质量，说明 GB 处理可作为采后维持血橙果实质量的有效方法（Habibi et al.，2022）。王懿（2021）发现，GB 处理能显著降低桃果实冷害，延缓相对电导率和 MDA 含量的增加，同时维持较高水平的 TSS、TA、游离氨基酸和有机酸的含量，抑制果实甜味和酸味的下降，维持较好的鲜味和丰富性。利用电子鼻和电子舌分析发现，对照组和处理组桃果实的味觉和气味上存在明显差异，经分析其中味觉差异的造成与桃果实中部分游离氨基酸、有机酸、TSS 和 TA 含量具有高度的相关性。表明 GB 处理能够有效减轻果实的冷害症状，并保持果实较好的风味品质。此外，GB 处理能够显著提高桃中精氨酸酶（Arginase，ARG）、精氨酸脱羧酶（Arginine Decarboxylase，ADC）、鸟氨酸脱羧酶（Ornithine Decarboxylase，ODC）和鸟氨酸转氨酶（Ornithine Transaminase，OAT）的活性，诱导 *PpARG*、*PpODC*、*PpADC* 和 *PpOAT* 基因的表达，促进多胺和脯氨酸的合成。同时，GB 处理能够降低二胺氧化酶（Diamine Oxidase，DAO）和多胺氧化酶（Polyamine Oxidase，PAO）的活性和 *PpPAO* 基因的表达，从而抑制多胺的分解代谢。表明 GB 处理可以促进精氨酸代谢，诱导多胺和脯氨酸的积累，保护了细胞膜的完整性，减轻桃果实的冷害。GB 处理提高了谷氨酸脱羧酶（Glutamic Acid Decarboxylase，GAD）和 γ-氨基丁酸转氨酶（γ Aminobutyric Acid Transaminase，GABA-T）活性，诱导了 *PpGAD*、*PpGABA-T* 和 *PpSSADH* 基因的表达，促进了 GABA 和琥珀酸的积累。表明外源 GB 处理通过调控 GABA 代谢促进 GABA 的合成，同时为 TCA 循环提供充足的琥珀酸底物，增强桃果实的耐冷性。GB 处理能够显著提高 APX、单脱氢抗坏血酸还原酶（Monodehydroascorbate Reductase，MDHAR）、脱氢抗坏血酸还原酶（Dehydroascorbate Reductase，DHAR）和谷胱甘肽还原酶（Glutathione Reductase，GR）的活性，诱导 *PpAPX*、*PpGR*、*PpMDHAR* 和 *PpDHAR* 基因的表达，维持桃果实中较高的抗坏血酸和谷胱甘肽含量以及 AsA/DHA 和 GSH/GSSG 比例，有效降低 H_2O_2 含量。表明外源 GB 能通过调节 AsA-GSH 循环系统，提高果实的抗氧化能力，从而减轻低温造成的氧化损伤。李灿婴等（2021）发现，用 5 mmol/L 甜菜碱处理可以有效抑制南果梨果实呼吸高峰，保持果肉硬度，延缓果实失重率增加、可滴定酸下降和可溶性固形物含量升高。与对照相比，5 mmol/L 甜菜碱处理显著地抑制了果实 PG、果胶甲基酯酶、多聚半乳糖醛酸反式消除酶及果胶甲基反式消除酶的活性。由此表明，甜菜碱处理通过抑制呼吸作用及果胶物质代谢相关酶活性延缓果实的软化，以此保持南果梨果实的贮藏品质。冷藏南果梨常温货架期易发生果皮褐变现象,研究发现,GB处理可有效减缓'南果梨'在0℃贮藏120 d后20℃货架期间果皮褐变的发生,具体表现为GB处理降低了果皮褐变指数和褐变发生率;脂过氧化程度低于对照,膜完整性保持较好;GB处理的果实中APX、CAT和SOD的活性和表达量均高于对照果实;GB处理果实中脯氨酸含量、脯氨酸合成关键酶活性和基因表达量均显著高于对照;此外,处理果实中脯氨酸含量、脯氨酸合成关键酶活性和基因表达量显著升高,包括鸟氨酸d-转氨酶(OAT)和Δ1-吡咯啉-5-羧酸合成酶(P5CS),这与褐变趋势一致。总之,GB处理可通过调节抗氧化酶和脯氨酸代谢有效缓解冷藏'南果梨'果皮褐变(Sun et al.,2020)。由于氧化应激和能量消耗,柑橘果实在采后贮藏期间的质量损失是常见的,Zheng et al.(2023)探讨了GB处理(10 mmol/L和20 mmol/L)对'黄果柑'果实采后常温贮藏期间品质及抗氧化活性的影响,发现10 mmol/L和20 mmol/L处理均有效降低了果实失重率和硬度损失,保持了可溶性固形物、可滴定酸和抗坏血酸含量。此外,GB处理显著提高了抗氧化酶活性,维持了较高的总酚和总黄酮水平,并导致H_2O_2积累缓慢,说明GB处理有利于提高'黄果柑'果实的耐贮性和延长货架期。这些结果表明,GB处理是调控采后果蔬理化性质的一种潜在的保存途径。

第四节　没食子酸丙酯对采后果实品质的影响

没食子酸丙酯(Proply Gallate,PG)为白色至浅褐色结晶粉末,或微乳白色针状结晶,无臭,微有苦味,水溶液无味。PG是一种常见的油溶性抗氧化剂和ROS清除剂(Morales et al.,2005),有清除自由基、阻止生物膜多种不饱和脂肪酸过氧化和保护抗氧化酶活性的作用(Zhang et al.,1996)。其抗氧化性优于BHT(二丁基羟基甲苯)及BHA(丁基羟基茴香醚),安全性高,是联合国粮农组织(FAO)和世界卫生组织(WTO)向全世界推行使用的优良油脂抗氧化剂之一。PG已被广泛应用于油脂、食品及化妆品中,如食用油脂、方便面、饼、果肉罐头等(李囡等,2007)。

甜瓜(*Cucumis melo* L.)属葫芦科黄瓜属一年生蔓性草本植物,我国是全世界最大的甜瓜产区(孟令波等,2001),现西北、华北、东北等地已成为了我国甜瓜生产的主产区(马德伟,1992)。甜瓜的果肉中富含碳水化合物、维生素、氨基酸、柠檬酸、果胶、葡萄糖、矿质元素等营养成分,水分多,味道可口,清爽芳香,更是解暑佳品,受到了广大消费者的喜爱(陈雷等,

1999）。甜瓜是呼吸跃变型果实，根据果实特征，可分为薄皮甜瓜和厚皮甜瓜2类（Sabato et al.，2015）。薄皮甜瓜较小，皮薄，可同果肉一起食用，但由于薄皮甜瓜采收期比较集中，而且正值高温多雨季节，加之薄皮甜瓜含糖量较高、水分含量大、生理代谢旺盛，采后若处理不当，则在贮运过程中极易受外界病菌侵染而发生腐烂变质，造成大量的经济损失（张润光等，2011）。因此，研究薄皮甜瓜的采后生理变化和贮藏技术，对于提高薄皮甜瓜的经济价值尤为重要。曹中权等（2016）以湖北省新洲区当地薄皮甜瓜为试材，采用不同浓度PG处理薄皮甜瓜，研究其采后生理变化和保鲜效果。发现不同浓度PG处理均具有较好的保鲜效果，其中0.50 mmol/L PG处理效果最好。该浓度处理后，甜瓜果实货架期得到了延长，果实腐烂率降低，果实贮藏期间的水分含量、可溶性糖、可滴定酸和维生素C含量均能保持较好的水平。PG处理组甜瓜的相对电导率、MDA含量、H_2O_2含量显著低于对照组，抗氧化酶类活性如CAT、SOD和POD均维持在较高水平，因此PG处理可显著减轻甜瓜果实的氧化损伤（曹中权等，2016）。说明PG是一种潜在的保鲜剂，对甜瓜采后贮藏保鲜具有不错的效果，其他呼吸跃变型果实的储藏保鲜研究可参考该研究结果。

为研究PG处理对新高梨果实软化与褐变的影响，在（0±0.5）℃，相对湿度90%~95%下，分析了不同质量分数的PG对冷藏期间新高梨果胶、PG活性、总酚和PPO活性的影响，发现PG减缓了新高梨的可溶性果胶含量的上升，抑制了PG酶活，进而减缓了果实软化衰老从而解释了为什么经过PG处理的果实贮藏效果要比未用PG处理过的果实要好。没食子酸丙酯能减缓新高梨总酚含量的下降，抑制PPO活性，这可能是因为，没食子酸丙酯抑制了果实的软化，进而减少果胶水解产生的氧自由基，并且抑制果实的软化，维持细胞组织的结构使底物、氧和PPO不能充分的接触，这些因素减缓了果实的褐变，抑制了PPO的活性，也有可能是因为没食子酸丙酯对PPO有直接的抑制作用，但机理有待于进一步研究。在整个贮藏过程中，PG酶的最大酶活出现时间总是要比PPO的要早。可以推断，在整个采后贮藏过程中，先是果实的软化衰老导致细胞组织的崩溃，进而产生大量氧自由基，使底物、氧和PPO充分的接触，进而加剧了果实的褐变（李汉良，2011）。外源没食子酸丙酯处理的龙眼可缓解果实的褐变现象，同时处理过的果实PLD，脂肪酶和LOX活性明显降低，说明该处理能通过抑制这3种膜脂代谢关键酶活性从而防止膜脂过氧化的发生，最终达到延缓龙眼果实褐变发生（Lin et al.，2013）。

库尔勒香梨（*Pyrus sinkiangensis* Yu）简称香梨（张钊等，1993），是新疆

巴音郭楞州和阿克苏地区的主栽果树品种（木合塔尔·扎热等，2012），因其香味浓郁，清甜可口，细嫩多汁，营养丰富而深受消费者喜爱，已成为新疆重要的出口果品之一（李慧民等，2008；高启明等，2005）。若香梨贮藏得当，可实现常年供应市场，但目前市场对香梨的品质要求越来越高，在贮藏过程中仍存在很多问题（刘琦等，2010）。为探究PG处理对库尔勒香梨贮藏期间品质的影响，采用PG处理香梨，通过定期检测其在贮藏期间的硬度、呼吸强度、细胞膜渗透率和相关保护酶等指标，发现PG溶液处理较好地保持了香梨果实硬度、可溶性固形物含量和总糖含量，抑制了香梨果实贮藏期呼吸强度，减缓果实细胞膜透性增加，对延缓果实内SOD、CAT活性下降和抑制POD活性有积极作用，说明PG处理可有效延缓库尔勒香梨贮藏期果实衰老，提高库尔勒香梨贮藏品质和保鲜效果（许娟等，2016）。

第五节　氯化钙处理对采后果实品质的影响

钙是植物细胞壁的重要组成部分，积极参与细胞壁结构的形成，维持细胞膜结构和功能的完整性（Aghdam et al.，2012）。氯化钙无毒、无臭、味微苦，是果实采后钙处理常用的试剂。采后外施氯化钙处理可以使钙离子与细胞壁中的果胶结合，提高细胞壁的刚性与支撑力，对保持果实采后硬度有着重要的作用。Manganaris et al.（2007）发现氯化钙处理可以抑制采后桃果实原果胶的降解，减缓水溶性果胶含量的增加，维持果实细胞壁完整。Guo et al.（2023）发现氯化钙处理显著提高荔枝果皮中钙离子和纤维素含量，降低了水溶性果胶含量、PG、β-gal和纤维素酶活性，加强了荔枝果实细胞壁结构的强度。Huang et al.（2023）研究发现外源氯化钙处理可通过提高果实中PME活性以及果胶酸钙含量导致枇杷果实果肉硬度在贮藏过程中不断增加。Silvia et al.（2019）研究发现氯化钙溶液处理草莓果实，提高了果实中PME的活性，并抑制了与果胶水解酶相对应的PG和β-gal的活性，此外钙处理还上调了草莓果胶代谢关键基因 *FaPME*1 的表达进而影响采后草莓果实的硬度。Jain et al.（2019）发现氯化钙处理收获的毛叶枣果实会导致纤维素酶，PG和PME的活性显著降低，从而提高毛叶枣果实的硬度。赵临强等（2021）发现采用不同浓度的钙处理均能抑制鲜切苹果的软化，提高其抗氧化活性，延缓鲜切苹果的品质劣变。潘家丽等（2023）通过对采后百香果采用氯化钙处理，发现能影响细胞壁中果胶和纤维素的含量，并通过延缓PG、PME活性的上升，调节细胞壁物质代谢来减缓百香果的软化速度，延长货架期。朱绍坤等

（2024）研究发现，喷施氯化钙后巨峰葡萄果实通过提升原果胶含量，降低可溶性果胶含量以及与细胞壁降解相关的 PG 和果胶裂解酶活性使葡萄的裂果率降低。寸丽芳（2023）研究发现外源喷施氯化钙处理能够显著增加骏枣果皮细胞壁纤维素、半纤维素、原果胶、共价结合型果胶和离子结合型果胶含量，降低果皮水溶性果胶含量及细胞壁果胶酶、纤维素酶活性和半纤维素酶活性，降低骏枣的裂果率。陈留勇等（2003）研究了不同温度下钙制剂对黄桃硬度的影响，研究表明，氯化钙可以延缓黄桃果实的后熟，保持果实的硬度。白琳等（2021）研究表明氯化钙处理可以通过抑制纤维素酶、PG 和 PME 活性延缓南果梨果实硬度的降低，有效延缓纤维素含量降低，抑制原果胶的降解和可溶性果胶含量的上升，有效延缓南果梨的采后软化进程。

此外，钙离子可抑制细胞水解酶活性，稳定细胞膜结构并影响膜上的调节蛋白，与钙调素一起诱导细胞内一系列生化反应，同时外钙离子可以作为植物细胞内的第二信使，具有调节细胞内部多种生理活动的功能（丘智晃等，2022），因此外施氯化钙处理可对采后果实多种生理活动起到调控的作用，进而影响采后果实的品质。Han et al.（2021）研究表明用氯化钙溶液处理的采后的苹果果实抑制了苹果果实可滴定酸和苹果酸含量的下降，增加了琥珀酸和草酸的含量，并通过影响相关基因的表达对苹果酸及 γ-氨基丁酸代谢途径产生影响。Wang et al.（2014）研究表明用氯化钙溶液处理采后甜樱桃果实，提高了总酚含量和总抗氧化能力，降低果实的呼吸速率以及膜脂过氧化，导致果实硬度和耐点蚀性增加，减少樱桃果实腐烂率和可滴定酸的损失。Madani et al.（2015）研究表明外源氯化钙处理采后番木瓜果实可提高其总抗氧化活性和总酚类化合物的含量，并降低了番木瓜炭疽病的发病率。孙思胜等（2022）研究表明 2%氯化钙处理可抑制'阳光玫瑰'葡萄果实落粒率和腐烂率的上升，延迟可滴定酸 TA 含量、维生素 C 含量和 TSS 含量的下降速率，在一定贮藏时间内可以保持'阳光玫瑰'葡萄果实的贮藏品质。刘雪艳等（2023）研究表明氯化钙处理小白杏可延缓果皮、果肉中叶绿素降解，抑制类胡萝卜的合成，维持果皮、果肉中 ATP、ADP 含量和能荷，延缓小白杏转色，从而延长小白杏采后货架期。李正国等（2000）通过对'白凤'水蜜桃施用氯化钙溶液，发现能显著抑制果实乙烯合成，从而推迟果实呼吸高峰的出现。张瑜瑜等（2022）研究表明氯化钙处理能有效降低采后蓝莓果实的腐烂率，减缓失重，抑制果实硬度和可溶性固形物含量的下降，显著提高蓝莓果实过氧化物酶活性，抑制 MDA 含量的上升，延缓类黄酮、总酚和花青素含量的下降，提高果实的贮藏品质。韩絮舟等（2020）研究表明氯化钙处理可降低红树莓果实在贮藏期间的呼吸强度，减缓果实硬度、可滴定酸含量、可溶性固形物含量、维生素 C 含量、总酚含量的下降与 MDA 含量的上升，抑制果实多酚氧化酶的活性，同时提高过氧化物酶活性及 CAT 活性，提高果实的贮藏品质。王玲利等（2018）研究表明氯化钙处理有助于皇冠梨果实硬度的维持，抑制果实可溶性固形物的流失和可滴定酸含量的下降，且随着钙浓度的增加效果越明显并可显著抑制维生素 C 含量的下降，提高果实的贮藏品质。王大伟等（2016）采用 1%、2%、3% 浓度的氯化钙处理新疆冬枣，发现不同浓度都能降低冬枣的腐烂率、失重率，保持果实硬度和营养物质含量，且 1% 浓度的保鲜效果最好。

  采后除了可直接外施氯化钙之外，还可通过氯化钙涂膜以及与其他保鲜方式相结合的方法来影响采后果实的相关生物化学变化，改善采后果实的品质。Kou et al.（2020）对采后的冬枣采用氯化钙和壳聚糖/纳米二氧化硅复合膜处理，抑制黄酮醇合成酶、二氢黄酮醇 4-还原酶和花青素合成酶的基因表达，促进亮蓝花青素还原酶的表达进而延缓了果实中花青素和槲皮素含量的增加，改善了采后果实品质的变化。Aiman et al.（2022）对采后的甜樱桃采用氯化钙壳聚糖膜处理可提高甜樱桃的总可溶性固形物和可滴定的酸度含量，从而保持了果实的品质。Chong et al.（2015）研究表明氯化钙结合壳聚糖可食性涂膜可减少鲜切蜜瓜 40% 的质量损失，使其增加 45%，整体颜色改变较少，抑制了微生物的生长，维持了可溶性果胶的完整性，进而延长了鲜切蜜瓜的保质期。Zhang et al.（2019）研究表明氯化钙与 1-MCP 配合使用能更有效地抑制新皇后甜瓜的呼吸强度、乙烯释放和原果胶水解，显著延缓果实软化，降低腐烂率，延长货架期。Elbagoury et al.（2020）研究表明用氯化钙与茉莉酸甲酯配合使用增加香蕉冷藏期间和成熟期果实中总酚类化合物的含量和总抗氧化活性，从而改善香蕉果实的采后品质和货架期，并能减轻冷藏和成熟温度下的冷害。李东等（2024）研究表明，2% 氯化钙浸泡+聚乙烯醇复合膜处理脆红李，可有效延缓其可滴定酸含量降低，减缓可溶性固形物含量上升，抑制脆红李的呼吸强度，抑制 PG 活性，降低 MDA 累积从而降低脆红李采后腐烂率，减缓果实后熟与衰老，延长货架期。张伟清等（2020）研究表明柠檬精油、壳聚糖、氯化钙形成的复合膜处理椪柑，可有效地提高果实维生素 C 和可滴定酸的含量，降低果实腐烂率、失重率、果皮相对电导率和果肉 MDA 含量，提高了果实的贮藏期。李自芹等（2023）研究表明氯化钙与 1-MCP 复合处理更好地抑制了采后西州蜜甜瓜果实的呼吸强度、原果胶物质的分解，延缓了西州蜜甜瓜的软化，保持了果实的贮藏品质。杨文慧等（2020）研究发现氯化钙和草酸处理可通过抑制香蕉果实 MDA 含量、细胞渗透率和 H_2O_2 含量的增加，促进游离脯氨酸的累积，增强 CAT、过氧化物酶、SOD 活性，提高采后香蕉的抗冷性，从而减轻低温冷害对香蕉的伤害。

因此，氯化钙可作为采后果实重要的外源处理试剂，可从调节细胞壁成分组成（如纤维素、果胶，半纤维素）以及细胞壁代谢相关酶（如半乳糖醛酸酶、果胶酶和纤维素酶）的活性影响果实采后硬度的变化。此外氯化钙处理还可通过影响采后果实体内可滴定酸，可溶性固形物，抗氧化物质的含量等多种生理指标进而改善采后果实的品质，达到保鲜的效果。氯化钙处理除了可直接喷洒使用外，还可进行复合涂膜或与其他保鲜手段相结合使用实现对采后果实品质的改善。

第六节 NO 处理对采后果实品质的影响

一氧化氮（Nitricoxide，NO）是近年来逐渐引起关注的一种气态信号分子，参与生物体内的多种代谢过程（章镇等，2012）。1996 年 Leshem et al. (1996) 首次报道 NO 可在植物体内合成，是果实成熟衰老的调控因子。已有较多证据表明，利用 NO 及其衍生物（如 N_2O）短时熏蒸或 NO 供体硝普钠（Sodium Nitroprusside，SNP）浸泡能够通过抑制果实贮藏过程的呼吸速率，并影响氧化物质代谢、糖代谢、膜脂过氧化和功能成分积累等，延缓果蔬组织的衰老进程、抑制乙烯的产生等作用来提高果蔬贮藏过程中对逆境胁迫的抵御能力，增强果蔬的保鲜效果延长果蔬货架期（Gu et al.，2014；Sahay et al.，2017；Wills et al.，2017；Zaharah et al.，2011；Beligni et al.，2000；Robertson et al.，1990；Lurie et al.，2005；Brummell et al.，2004；焦彩凤等，2019）。近年来，NO 气体已广泛用于果蔬采后贮藏保鲜过程中（Xu et al.，2012；Yang et al.，2011）。NO 的供体物质 SNP 能够通过抑制果实呼吸作用来推迟其后熟软化进程（Tran et al.，2015），如 SNP 能够通过延缓桃（Koushesh et al.，2017）、李（Sharma et al.，2015）、樱桃（Rabiei et al.，2019）等果实的后熟软化进程，减少病原菌侵染等，提高水果贮藏过程中对逆境胁迫的抵御能力，进一步改善果实采后贮藏品质，延长果实保质期（Hu et al.，2019）。此外，外源 NO 处理还能有效延缓草莓（Wills et al.，2000；朱树华等，2005）、番茄（李翠丹等，2013）、番木瓜（郭芹等，2011）、桃（Zhu et al.，2019）以及猕猴桃（Zheng et al.，2017；朱树华等，2009）衰老。NO 处理采后草莓可提高其贮藏期间可溶性固形物和可滴定酸的含量，降低维生素 C 含量的损失，进而推迟推迟草莓果实固酸比、维生素 C 和可溶性蛋白含量下降（朱树华等，2005）Sharma et al.（2015）发现，0.5 mmol/L 外源 NO 处理能促进李果实贮藏期间酚类物质含量的增加。Zheng et al.（2017）的研究表明，猕猴桃贮藏期间果实的总酚类物质、黄酮类化合物含量在外源 200 μmol/L NO 处理下也有所升高。经 SNP 处理后，华冠油菜的黄化率显著降低（陈海荣等，2007）。王云香等（2018）研究发现，NO 处理可通过降低 MDA 含量，维持 TSS、维生素 C 和叶绿素含量，提高 POD 和 CAT 活性，较好地维持黄瓜采后品质和营养价值。

百香果（*Passi flora coerulea* L.）也叫西番莲、鸡蛋果，因含有多种水果香气而得名，口感酸甜，风味宜人（Charan et al.，2017）。百香果富含氨基酸、蛋白质、糖、维生素等多种营养物质（余东等，2004），具有抗氧化（Silva et al.，2013）、抗肿瘤（Silva et al.，2012）、抗焦虑（Deng et al.，2010）等功效，还能清热降火、止咳化痰（Dhawan et al.，2002）。百香果成属于高温季节，且属于呼吸跃变型果实，采后 2~3 d 会发生呼吸跃变，随后表皮出现皱缩现象，果肉液化且有异味，影响品质和价值（Pruthi.，1963）。王红林等（2021）用 50 μmol/L、100 μmol/L、200 μmol/L NO 处理百香果，发现 NO 处理可减少贮藏期间百香果的腐烂率和失重率，延缓了色泽 L、a、b 值的降低，维持了维生素 C 和可溶性蛋白质含量以及贮藏后期的糖酸比水平，促进了总酚和类黄酮含量的增加，其中中间浓度（100 μmol/L）效果更佳。常雪花等（2019）和 Zhao et al.（2020）研究了 NO 熏蒸处理对冬枣的保鲜效果，结果表明 150 μL/L NO 处理后在（0±1）℃条件下贮藏效果最好。豇豆（*Vigna sinensis*）又称粉豆、长荚豆，味道鲜美，营养丰富，但采后常温贮藏易失水（王利斌等，2013；范林林等，2015）。将豇豆用 0.2 mmol/L、0.3 mmol/L、0.4 mmol/L SNP 溶液浸泡 10 min，在 8 ℃ 的恒温恒湿箱中贮藏 10 d，每隔 2 d 监测生理生化指标。发现 SNP 处理能有效降低豇豆的水分流失和抑制乙烯释放量，维持可溶性固形物、维生素 C、叶绿素含量，抑制相对电导率上升，维持细胞膜稳定性，降低 PPO 活性，提高 POD 和 CAT 活性，说明 SNP 处理可以较好地维持豇豆的贮藏品质，延长贮藏时间（范林林等，2015）。程顺昌等（2005）发现，适当浓度 NO 处理可延缓青椒维生素 C 降解，保持叶绿素含量。用 20 μL/L NO 熏蒸处理草莓果实，可抑制草莓果实腐烂，延缓 TSS 含量下降，降低膜脂质过氧化程度，维持果实品质；还能提高果实中 PAL、肉桂酸羧化酶（Cinnamate－4－hydroxylase，C4H）、查尔酮异构酶（Chalcone isomerase，CHI）、4-香豆酸 CoA 连接酶（4-coumarate A ligase，4CL）等苯丙烷类代谢相关酶活性，保持果实较高的总酚、花色苷、类黄酮含量，维持较高的抗氧化活性；此外，该处理还通过调控果实抗氧化酶类如过 CAT、SOD 和 APX 等活性，诱导草莓贮藏前期 H_2O_2 的产生和抑制后期果实 H_2O_2 含量的上升；通过抑制 LOX 活性的上升，降低果实中 MDA 的积累，说明，NO 处理调控采后草莓果实 ROS 代谢、苯丙烷类代谢以及病程相关蛋白活性来延缓果实成熟衰老、抑制果实腐烂（黄玉平，2016）。红树莓（*Rubus ideaus* L.）又被称为悬钩子、覆盆子、木莓等，是蔷薇科、悬钩子属的木本植物，为多年生小灌木（Turmanidze et al.，2017）。红树莓果实色泽宜人，果香浓郁，酸甜可口，富含总酚、类黄酮、花青素等多种有益活性成分以及人体所必需的八种氨基酸，营养价值极高，深受消费者喜爱（王建勋等，2006；赵文琦等，2007；Gonzalez et al.，2011；朱雪静，2018）。成熟红树莓果皮薄，组织娇嫩，呼吸速率高，迅速失去水分和重量，因此采后不耐贮藏，保质期非常短（杨铨珍等，1992）。新鲜采摘的树莓应在 2~3 d 内食用。Shi et al.（2019）分别用 5 μmol/L、15 μmol/L、30 μmol/L NO 处理树莓果实，发现 NO 能维持较高的花青素、芦丁、总酚和总黄酮含量，减少超氧阴离子和羟基自由基的释放；提高红树莓贮藏过程中可溶性糖代谢相关酶活性，调控可溶性糖的组成。且 15 μmol/L NO 比其他浓度更适合提高红树莓的抗氧化能力，增强可溶性糖代谢，进而维持红树莓的贮藏品质。王俊文（2020）发现 NO 处理可提高树莓的抗氧化系统、苯丙烷和花青素代谢，从而提升采后树莓在贮藏期间的品质。

NO 处理能提高冷藏秋葵 SOD、过氧化物酶和 CAT 的活性，增强对氧自由基的清除能力（杨小兰，2020）。Ren et al.（2023）研究结果表明，NO 处理可龙眼增强几丁质酶（Chitinase，CHI）、β-1,3-葡聚糖酶、苯丙氨酸氨裂解酶、多酚氧化酶的活性，提高总酚含量，还抑制了细胞壁修饰酶的活性，保持了细胞壁组分，较好地维持了龙眼的品质。已有研究发现，利用 NO 处理果蔬能起到保鲜的效果，其作用机制可能与抑制乙烯产生有关（Veiga et al.，2024；Leshem et al.，2000；朱丽琴等，2013；周春丽等，2011）。在菠菜（周春丽等，2012）、油菜（陈海荣等，2007）、桃（Franciscob et al.，2008）、香蕉（Cheng et al.，2009）、芒果（杨杨等，2013）、猕猴桃（朱树华等，2009）、番茄（张少颖等，2005）、鲜莲子（王瑶等，2019）、番木瓜（郭芹等，2013）等果蔬中均证实，适当浓度的 NO 处理可通过降低乙烯释放速率来抑制呼吸作用，从而对果蔬达到保鲜的目的。

第六章 生物保鲜处理对采后果实品质的影响

生物保鲜通常是采用拮抗菌抑制致病菌的生长或利用天然植物提取物的杀菌性来预防果蔬腐烂变质（郭慧静等，2023）。

第一节 褪黑素处理对采后果实保鲜的影响

作者用褪黑素处理了采后低温贮藏南果梨，并探究了褪黑素处理对果皮褐变影响（孙华军，2020），因此该部分内容以褪黑素对冷藏南果梨果皮褐变的影响为主。

一、褪黑素简介

褪黑素（MT）是一种低分子量吲哚胺类激素，也是一种重要的信号分子，参与植物的多种生物过程，如种子萌发、果实成熟和衰老等，在植物、动物和微生物中广泛存在（Aghdam et al.，2017；Aghdam et al.，2019a；Hardeland，2016；Sun et al.，2015；Hernández et al.，2015；Manchester et al.，2000；Burkhardt et al.，2001；Paredes et al.，2009；Posmyk et al.，2009）。MT也是一种强大的自由基清除剂和抗氧化剂，可以增强植物的抗逆性（Tan et al.，1993），响应各种胁迫，如冷（Bajwa et al.，2014）、盐（Li et al.，2012），干旱（Wang et al.，2017），氧化损伤（Martinez et al.，2018）和营养缺陷等（Kobyli'nska et al.，2018）。MT于1958年首次从牛松果腺中分离鉴定，因为它可以减轻肤色并抑制黑素细胞刺激素，因此有人认为这种物质被称为MT（Lerner et al.，1958）。在MT被分离后的40年，关于MT的研究主要集中在动物身上。研究表明，MT在调节抗氧化酶活性（Pieri et al.，1994；Rodriguez et al.，2004），昼夜节律（Brainard et al.，2001），冠心病（Brugger et al.，1995），阿尔茨海默病（Cardinali et al.，2002），身体状况和情绪状态

(Dollins et al., 1994）中发挥了关键作用。1995年，MT在植物中被两个团队发现（Dubbels et al., 1995；Hattori et al., 1995），之后，在不同的植物物种中均检测到了MT的存在，并且MT在植物生理过程中的作用引起了广泛的关注。

二、褪黑素在果实采后贮藏中的应用

MT处理具有积极调节ROS相关抗氧化基因和CBF表达的能力，从而提高了拟南芥对寒冷胁迫的耐受性（Bajwa et al., 2014）。Aghdam et al.（2019b）报道，外源MT处理通过提高鲜切花中苯丙氨酸解氨酶（PAL）活性，降低PPO活性，积累更多的脯氨酸和酚类化合物，提高DPPH的清除能力，并降低有害H_2O_2浓度，从而降低膜的通透性和膜脂质过氧化作用。MT处理可通过提高6-磷酸葡萄糖酸脱氢酶（G6PDH），草酸脱氢酶（SKDH）和PAL活性来降低桃果实中的冷害，该处理还抑制了PPO活性，防止膜脂过氧化发生，并保持较高的不饱和脂肪酸与饱和脂肪酸比率（Gao et al., 2018）。外源MT可以通过提高CCO和SDH的活性从而增加冷藏过程中番茄体内ATP含量，并通过诱导脂肪酸去饱和酶（FAD3和FAD7）的表达，降低PLD和LOX活性及转录水平，从而提高亚油酸和亚麻酸积累，降低硬脂酸和棕榈酸积累，使不饱和脂肪酸与饱和脂肪酸比值（U/S）提高，最终缓解了番茄冷害的发生（Jannatizadeh et al., 2019）。Aghdam et al.（2019a）报道，外源MT处理通过上调ZAT2/6/12表达，激发精氨酸路径活性，从而提高内源多胺、脯氨酸和一氧化氮积累，并诱导CBF1的表达，提高番茄果实的耐冷性，从而降低其冷害。血清素N-乙酰基转移酶（SNAT）和N-乙酰基5-羟色胺甲基转移酶（ASMT）是MT生物合成的两个关键酶，抑制水稻内源SNAT和ASMT后发现水稻内源MT含量减少，缺乏SNAT的水稻出现幼苗生长延缓的现象，通过外源MT处理后该现象可以部分恢复，表明MT在幼苗生长中起作用，此外，这种水稻对各种非生物胁迫（包括盐和寒冷）更加敏感（Byeon et al., 2016）。MT处理几乎完全缓解了冷应激引起的胡萝卜细胞质膜的收缩和破坏（Lei et al., 2004）。外源MT处理能提高抗氧化酶APX活性，提高抗氧化物质抗坏血酸和谷胱甘肽含量，延缓淀粉降解；也能提高抗氧化酶SOD和CAT活性，减少·O^{2-}和H_2O_2积累，降低了低温对猕猴桃细胞膜的伤害，延缓了相对电导率和MDA的升高，有效延迟了冷害发生时间，降低了冷害率和冷害指数（胡苗，2018）。MT处理能通过维持黄瓜贮藏过程中的维生素C、叶绿素、可溶性蛋白和可滴定酸含量从而能保持黄瓜的品质；进一步研究发现褪黑素处理的果实抗氧化酶活性（如 SOD、CAT 和 APX）较高，相对电导率、MDA 和 H_2O_2 含量降低，冷害指数和失重率显著低于对照果实（辛丹丹，2017）。为研究褪黑素（MT）诱导采后黄瓜果实耐冷性的机制，采用 100 μmol/L MT 处理黄瓜果实，在 4 ℃、相对湿度 90% 条件下贮藏 15 d。与对照相比，MT 处理减轻了黄瓜的冷害，降低了电解质渗漏率，提高了果实硬度。MT 处理的果实在贮藏条件下表现出较高的叶绿素含量，叶绿素酶活性受到抑制。MT 处理提高了精氨酸脱羧酶（ADC）和鸟氨酸脱羧酶（ODC）酶活性。此外，黄瓜 ADC（CsADC）和黄瓜 ODC（Cs ODC）基因的表达增强导致多胺含量的积累。同样，脯氨酸水平在处理的果实中表现出更高的水平。同时，脯氨酸合成酶 Δ1-吡咯啉-5-羧酸合成酶（P5CS）和鸟氨酸氨基转移酶（OAT）活性显著增加，而脯氨酸分解代谢酶脯氨酸脱氢酶（PDH）的活性受到抑制。此外，MT 诱导 C. sativus OAT（CsOAT）和 C. sativus P5CS（CsP5CS）基因表达。MT 处理的黄瓜果实也表现出更高的 γ-氨基丁酸（GABA）含量，并进一步增强了 GABA 转氨酶（GABA-T）和谷氨酸脱羧酶（GAD）酶活性，促进了黄瓜 GAD（Cs GAD）基因表达，表明 MT 处理增强了采后黄瓜果实的耐冷性，且与多胺、脯氨酸和 γ-氨基丁酸的调控有关（Madebo et al.，2021）。外源 MT 处理也能提高采后桃果实冷藏期间的耐冷性。Cao et al.（2016）发现，外源 MT 处理显著提高了采后桃果实冷藏期间的多胺和 GABA 含量，提升了 PpP5CS 和 PpOAT 的表达，降低了 PpPDH 的转录，导致脯氨酸含量增加，激活了潜在的正反馈机制，并增强了桃果实的耐寒性，这些结果表明，MT 在冷胁迫下能在植物细胞中发挥抗凋亡作用。

三、褪黑素对冷藏南果梨褐变的延缓作用研究

南果梨酸甜适口、果肉细腻而多汁、香气浓郁而独特，而且含有多种脂肪酸、黄酮、氨基酸、丰富的粗蛋白、粗纤维和可溶性糖等，深受消费者喜爱（董萍，2011；侯冬岩等，2005）。在农业农村部重点区域发展规划中，南果梨被列为重点发展的特色梨品种之一（姜巍，2015）。近年来，南果梨的栽培面积不断扩大，在辽宁省鞍山市和辽阳市的栽培面积已达 80 多万亩，是辽宁省的第二大果树品种（庄晓红等，2008）。

南果梨的采收期通常在 9 月上中旬（纪淑娟等，2012；王学密等，2008），它是呼吸跃变型果实（吴震等，1997），采收后在常温条件下自然后熟 15 d 左右达到呼吸跃变峰和乙烯释放高峰，此时品质最佳，之后随着衰老的加剧，果心开始褐变，严重时果肉也发生褐变，最后腐烂，失去了其特有的

商品品质（于天颖等，2010）。针对这一问题，多年来研究者们致力于南果梨冷藏保鲜技术研究，他们发现冷藏能抑制果实呼吸，延缓果实后熟衰老，防止微生物的侵害，延长南果梨的贮藏期至6个月，解决了南果梨集中销售的问题，大大提高了果农的经济效益（卜庆状，2012；宜景宏等，2003）。然而实践中发现，经过长期冷藏的南果梨在常温货架期间果皮很容易出现褐变症状，果实的香气也明显变淡，影响果实品质，降低了其商品价值（程顺昌，2013；张丽萍，2013；周鑫，2015；盛蕾，2016；Zhou et al.，2015a）。

为了探索褪黑素处理对冷藏南果梨果皮褐变的调控作用，在果实入库前，采用褪黑素对南果梨进行处理，通过观察冷藏后果实褐变的发生情况及果皮组织细胞超微结构的变化，测定果皮组织细胞膜透性和MDA含量，果皮组织脂肪酸和磷脂组分含量、膜脂代谢关键酶活性以及膜脂降解关键基因 $PuPLDβ1$ 和转录因子 PuMYB21 和 PuMYB54 相关基因表达量，研究贮前褪黑素处理对冷藏南果梨果皮褐变的调控作用，为生产上缓解冷藏南果梨品质劣变问题提供理论依据和技术支撑。

（一）褪黑素处理对冷藏南果梨果皮褐变的影响

褐变率和褐变指数可以反映果实褐变的发生情况。如图6-1所示，冷藏120 d的南果梨在出库当天，对照和褪黑素处理组果实均未出现褐变现象，但随着常温货架期的延长，对照果实在第6 d果皮开始出现明显的褐变症状，而且随着货架期的延长，褐变率和褐变指数快速升高，而MT处理的果实在第9 d才出现果皮褐变，并且褐变率和褐变指数均显著低于同期对照果实。至货架第12 d时，MT处理果实的褐变率和褐变指数分别为33%和25%，分别比对照

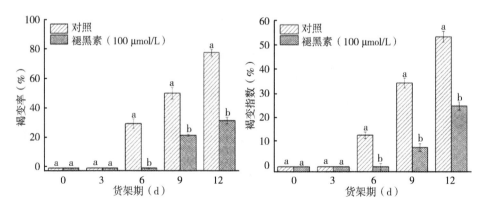

图6-1 不同处理南果梨冷藏后常温货架期果皮褐变率和褐变指数变化

低 2.36 倍和 2.14 倍。由此可见，MT 处理有效抑制了冷藏南果梨果皮褐变的发生及其发展。

（二）褪黑素处理对冷藏南果梨细胞超微结构的影响

如图 6-2 所示，两种处理果实冷藏 120 d 后转入常温货架期第 6 d 时细胞超微结构有差异。褪黑素处理后的果实细胞器排列紧密，叶绿体和线粒体双层膜结构清晰可见，线粒体中脊排列有序，结构完整，细胞膜与细胞质贴合紧致。然而，对照果实果皮的细胞器数量变少，叶绿体和线粒体双层膜结构模糊，出现退化现象，并且叶绿体内的嗜锇颗粒明显多于褪黑素处理的果实。对照果实虽然没有出现质壁分离现象，但是细胞膜结构已经受到破坏（图 6-2B2），表明褪黑素处理能提高细胞内部结构完整性。

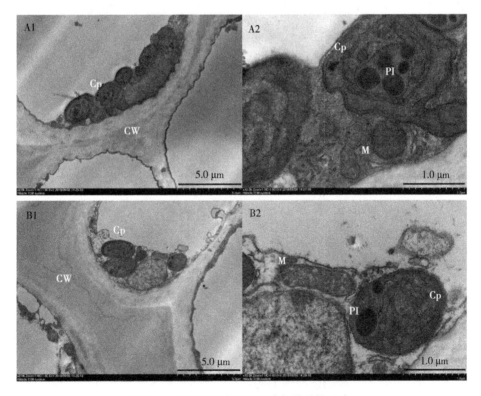

图 6-2 不同处理南果梨细胞超微结构观察

冷藏南果梨在常温货架第 6 d 的果皮组织细胞超微结构。A1，A2：MT 处理果实；B1，B2：对照果实。CW：细胞壁；Cp：叶绿体；Pl：质体小球；M：线粒体。

(三) 褪黑素处理对冷藏南果梨细胞膜透性和丙二醛含量的影响

相对电导率常用来反应果实细胞膜的透性。由图 6-3 可知，出库当天，两种处理果实的果皮相对电导率均较低，二者之间无显著差异，说明 MT 处理对冷藏过程中果实细胞膜透性的影响不显著。转入常温货架期后，随着果实的后熟过程，两种处理果实的相对电导率均呈逐渐上升趋势，尤其是在货架的前 3 天，但是，经 MT 处理的果实相对电导率的上升幅度远小于对照果实，整个货架期 MT 处理果实的相对电导率始终显著低于同期对照果实，说明 MT 处理有效保护了冷藏南果梨细胞膜的完整性。

MDA 是膜脂过氧化的产物，其含量可以反映膜脂过氧化的程度。与相对电导率的变化趋势相似，出库时 MT 处理与对照果实的果皮 MDA 含量无显著差异，随着常温货架的延长二者均呈上升趋势。但是与对照相比，MT 处理果实 MDA 含量始终处于较低水平，表明 MT 处理有效抑制了冷藏南果梨的膜脂过氧化作用。

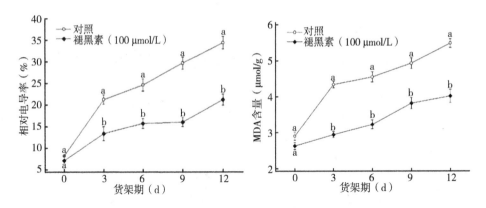

图 6-3 不同处理南果梨冷藏后常温货架期相对电导率和丙二醛含量变化

(四) 褪黑素处理对冷藏南果梨果皮组织脂肪酸的影响

在南果梨果皮中检测到了 3 种不饱和脂肪酸，包括油酸、亚油酸和亚麻酸，以及 2 种饱和脂肪酸，包括硬脂酸和棕榈酸。如图 6-4 所示，出库当天，除亚油酸外，对照和褪黑素处理果实的其余 2 种不饱和脂肪酸和 2 种饱和脂肪酸相对含量均表现出一定差异，说明褪黑素处理在冷藏过程中对南果梨中不同脂肪酸组分的含量已产生影响。转入常温货架期后，随着常温货架的延长，油酸相对含量呈上升趋势，而亚油酸和亚麻酸相对含量呈下降趋势，两种不饱和脂肪酸含量均呈波动变化。与对照相比，褪黑素处理果实在整个货架期内亚油

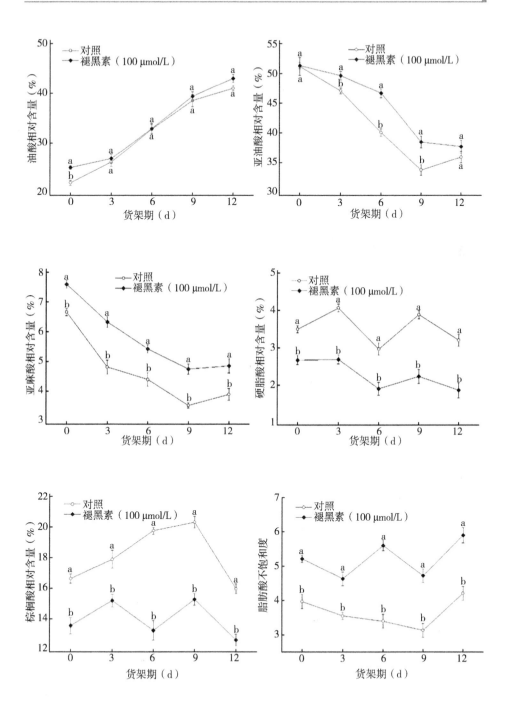

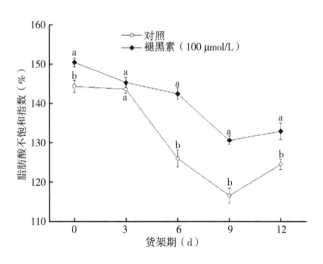

图 6-4　不同处理南果梨冷藏后常温货架期脂肪酸相对含量变化

酸（常温货架第 12 d 除外）和亚麻酸相对含量均显著高于对照果实，褪黑素处理对亚油酸和亚麻酸相对含量变化的影响较为明显。对照和褪黑素处理果实的硬脂酸相对含量呈波动变化趋势，均在常温货架期第 3 d 时达到峰值，但是褪黑素处理果实的硬脂酸相对含量始终低于同期对照果实。与硬脂酸类似，褪黑素处理果实的棕榈酸相对含量也呈波动变化，而对照组果实的棕榈酸相对含量呈先升高后降低的变化趋势，在常温货架期第 9 d 时达到峰值。对照果实中硬脂酸和棕榈酸的相对含量一直处于较高水平的波动变化，而褪黑素处理果实硬脂酸和棕榈酸相对含量始终低于同期对照果实。由此可见，褪黑素处理有效抑制了冷藏南果梨果实中不饱和脂肪酸含量的下降和饱和脂肪酸含量的升高。

脂肪酸不饱和度和脂肪酸不饱和指数通常用来反映脂肪酸的不饱和程度。出库当天，经褪黑素处理的果实脂肪酸不饱和度和不饱和指数均显著高于对照果实。随着常温货架的延长，褪黑素处理果实脂肪酸不饱和度呈波动变化趋势，对照果实的脂肪酸不饱和度先下降后升高，但是褪黑素处理果实的脂肪酸不饱和度在整个常温货架期均显著高于同期对照果实。尽管处理与对照果实在货架的前 9 d 不饱和指数都呈下降趋势，而后又升高，但是经褪黑素处理的果实不饱和指数的下降幅度远小于对照。说明褪黑素处理能一定程度维持果实脂肪酸的不饱和程度。

（五）褪黑素处理对冷藏南果梨膜脂组分的影响

在南果梨样品中共检测到了 6 种磷脂，包括：PA、PG、PS、PI、PC 和

PE；两种溶血磷脂，包括：LPC 和 LPE；以及两种糖脂，包括：MGDG 和 DGDG（图 6-5）。可以看出，不论对照还是褪黑素处理果实，MGDG 和 DGDG 含量最高，PI 次之。褪黑素处理的果实 DGDG、PA 和 LPC 相对含量显著低于对照，而 PC、PE 和 MGDG 相对含量显著高于对照。其他脂类分子，如 PG、PS、PI、LPE 相对含量在对照和处理果实之间没有显著差异。

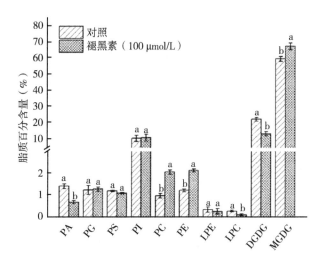

图 6-5 不同处理南果梨冷藏后常温货架期脂质相对含量的变化

（六）褪黑素处理对冷藏南果梨 PLD 活性和相关基因表达量的影响

PLD 是植物组织膜脂代谢的关键酶，可催化膜脂主要成分磷脂的降解。由图 6-6 可以看出，出库当天，两种处理果实的 PLD 活性没有显著差异。随着常温货架期的延长，对照果实 PLD 活性呈先升高后降低的变化趋势，第 9 d 时达到峰值，随后骤然下降。与对照果实相反，褪黑素处理果实的 PLD 活性在常温货架期的第 3 d 开始急剧下降，至第 6 d 时达到最低值，然后快速升高，然而在整个常温货架期，褪黑素处理果实的 PLD 活性显著低于同期对照果实。与 PLD 活性类似，两种处理果实的 PuPLDβ1 相对表达量在出库当天没有显著差异。对照果实 PuPLDβ1 相对表达量在整个常温货架期呈上升趋势，并且在货架中后期上升的更明显。而褪黑素处理果实的 PuPLDβ1 相对表达量在整个常温货架期一直处于很低水平。表明褪黑素处理在冷藏过程中对 PLD 活性无明显的影响，但是在冷藏后常温货架期能显著抑制 PLD 活性的升高。

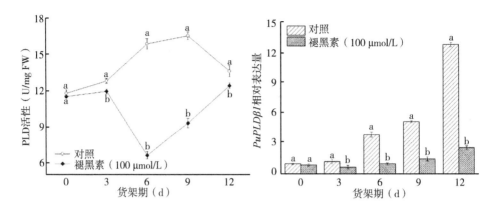

图 6-6　不同处理南果梨冷藏后常温货架期 PLD 活性及表达水平变化

（七）褪黑素处理对冷藏南果梨 LOX 活性和相关基因表达量的影响

冷藏 120 d 的南果梨出库时，经褪黑素处理的果实 LOX 活性显著低于对照（图 6-7），说明该处理对冷藏过程中果实的 LOX 活性有一定的影响。随着常温货架期的延长，两种处理果实的 LOX 活性均呈先升高后降低的变化趋势，但是，褪黑素处理的果实 LOX 活性的上升幅度远小于对照果实，高峰时，褪黑素处理果实的 LOX 活性比对照果实低 36.04%，而且，处理 LOX 活性高峰延晚 3 d 出现，由此可见，褪黑素处理可在一定程度上抑制 LOX 活性。与 LOX 活性不同，两种处理果实的 PuLOX1d 相对表达量在出库当天并无显著差

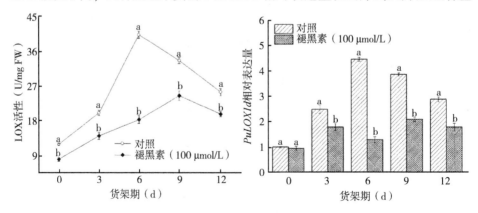

图 6-7　不同处理南果梨冷藏后常温货架期 LOX 活性及表达水平变化

异，但是在随后的常温货架期间，对照果实 PuLOX1d 相对表达量剧烈升高，货架第 6 d 时达到高峰，然后快速下降，而褪黑素处理果实 PuLOX1d 相对表达量在较低水平波动变化，并未出现明显的高峰。表明褪黑素处理能显著抑制冷藏后常温后熟过程果实的 PuLOX1d 的表达。

（八）褪黑素处理对冷藏南果梨脂肪酶活性和相关基因表达量的影响

如图 6-8 所示，尽管两种处理果实在冷藏后常温货架期间脂肪酶活性均呈先下降后上升的变化趋势，但是，在货架的初期褪黑素处理果实的脂肪酶活性即已显著低于对照果实，而且随着货架期的延长，对照果实脂肪酶活性下降至第 6 d 后骤然升高，而处理果实脂肪酶活性平稳下降至第 9 d 后略有升高，除第 6 d 外，处理果实的脂肪酶活性在货架期显著低于同期对照果实。与脂肪酶活性略有不同，两种处理果实相对表达量呈现相似的变化趋势，均在货架前期缓慢下降，第 6 d 之后快速升高，但是，无论是在出库当天还是在之后的常温货架期，褪黑素处理果实的 Pulipase 相对表达量均显著低于对照果实，由此可见，贮藏之前褪黑素处理对冷藏过程及冷藏后常温货架期脂肪酶活性及相关基因的表达均有一定的抑制作用。

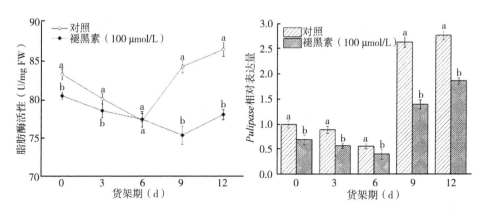

图 6-8 不同处理南果梨冷藏后常温货架期脂肪酶活性及表达水平变化

（九）褪黑素处理对冷藏南果梨抗氧化酶活性和相关基因表达量的影响

SOD、CAT、APX 和 POD 是果实中主要的抗氧化酶，可保护膜脂组分，防止发生膜脂过氧化。冷藏 120 d 的南果梨出库当天，褪黑素处理果实 SOD 和

APX活性均显著高于对照果实，两种处理果实CAT和POD活性无明显差异（图6-9）。随着常温货架期的延长，对照果实SOD和APX活性先快速下降，然后缓慢升高，再快速降至较低水平，而经褪黑素处理的果实SOD和APX活性在整个货架期比较平稳，而且始终显著高于对照果实。与SOD和APX活性的变化趋势不同，对照果实在货架的前3 d CAT活性缓慢下降，然后缓慢上升，至货架第9 d后骤然下降，而处理果实在货架的前3 d CAT活性缓慢上升，然后快速升高至第6 d，之后一直稳定在较高水平。对照果实POD活性在货架前中期快速升高，之后迅速下降，而处理果实POD活性在整个货架期呈直线上升趋势。除初始阶段外，在整个常温货架期处理果实CAT和POD活性均显著高于对照果实。

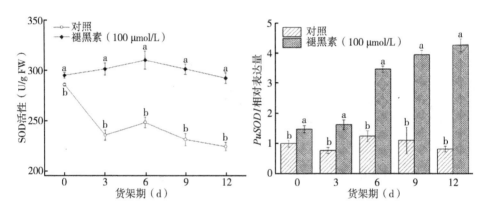

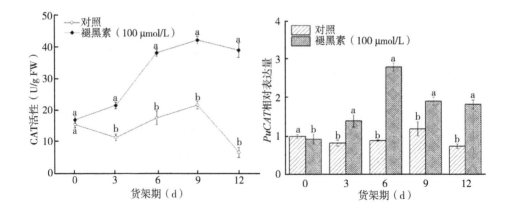

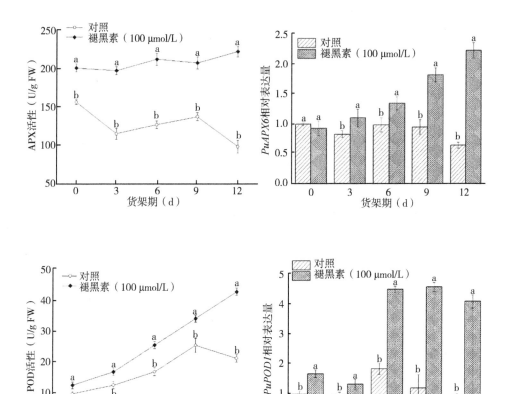

图6-9 不同处理南果梨冷藏后常温货架期抗氧化酶活性和表达水平变化

在整个常温货架期间，对照果实 PuSOD1、PuCAT、PuAPX6 和 PuPOD1 相对表达量在较低水平呈波动变化，而褪黑素处理果实 PuSOD1 和 PuAPX6 的相对表达量却随着货架期的延长而逐渐升高，尤其是 PuSOD1，在货架第 6 d 时骤升至对照的 2.79 倍，之后稳定在高水平。处理果实 PuCAT 相对表达量在货架第 3 d 之后剧烈升高，至第 6 d 时达到高峰，此时，其表达量为对照果实的 3.16 倍，然后逐渐下降，但是，在货架后期处理果实 PuCAT 相对表达量仍显著高于同期对照果实。对照果实 PuPOD1 相对表达量在货架前期较为平稳，第 6 d 时剧烈升高，随后骤然下降，处理果实 PuPOD1 的相对表达量在货架前期不高，货架中后期剧烈升高至对照果实的 2 倍多，并且在整个常温货架期处理果实的 PuSOD1、PuCAT、PuAPX6 和 PuPOD1 相对表达量显著高于同期对照果实。由此可见，褪黑素处理可有效激活冷藏南果梨果实的抗氧化酶系统，这对于保护细胞膜脂组分的稳定性具有重要意义。

（十）褪黑素处理对冷藏南果梨 PuMYB21 和 PuMYB54 表达水平的影响

前期研究表明，长期低温胁迫能诱导 PuMYB21 和 PuMYB54 表达量增加，从而激活其下游基因 PuPLDβ1 的表达，促进膜脂降解，破坏细胞膜结构，促使果皮褐变发生。为此，本试验测定了褪黑素处理对这两个 MYB 转录因子家族成员表达水平的影响，如图 6-10 所示，在货架的前 9 d 对照果实 PuMYB21 相对表达量随着冷藏后常温货架期的延长呈梯度升高，之后稳定在较高水平，褪黑素处理的果实 PuMYB21 相对表达量虽然呈现类似的变化趋势，但是，其上升速率与上升幅度远小于对照果实，至货架第 9 d 时，其表达量仅为对照果实的 40.09%。同样，处理与对照果实 PuMYB54 相对表达量的总体变化趋势也是相似的，均在货架的前 9 d 逐渐升高而后快速下降，但经褪黑素处理的果实无论是在出库当天还是在随后的常温货架期间 PuMYB54 相对表达量始终显著低于对照果实。可见，该处理对 PuMYB21 和 PuMYB54 的表达有一定的抑制作用。

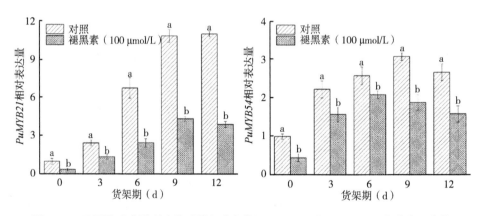

图 6-10 不同处理南果梨冷藏后常温货架期 PuMYB21 和 PuMYB54 表达水平变化

前期研究结果表明，长期冷藏的南果梨出库后在常温货架期间会出现果皮褐变现象，并伴随着膜脂代谢失调和果皮细胞膜系统破坏（Sun et al., 2020a）。通过系统研究，还证明了 PuPLDβ1 是低温胁迫下膜脂组分与功能变化的关键基因，转录因子 PuMYB21 和 PuMYB54 能协同调控 PuPLDβ1 的上调表达,从而应答低温胁迫,促进冷藏南果梨果皮褐变的发生(Sun et al., 2020b)。基于上述机理研究结果,探索相应的调控技术极其重要。关于缓解低温伤害的方法已有一些报道,包括物理方法如程序降温和间歇升温,以及化学方法如甜菜碱、1-MCP、一氧化氮、精氨酸、ABA 和 MT。MT 是一种吲哚胺激素,也是帮助植物抵抗生物和非生物胁迫的重要渗透调节物质,近年来关于其在果实采后的应用及机理方面的研究较多。MT 处理能降低冷藏过程中桃果实的氧化损伤和脂质过氧化发生,以提高果实耐冷性并缓解冷害现象(Cao et al., 2016)。最近的研究表明,MT 处理能增强甜樱桃的抗氧化能力,从而延缓果实采后衰老进程(Wang et al., 2019b)。Miranda et al.(2020)发现,MT 处理过的呼吸跃变和非呼吸跃变甜樱桃 MDA 含量显著降低,这种变化与编码 CAT-SOD 和 AsA-GSH 循环中整体基因的转录水平诱导相关,说明 MT 处理通过触发抗氧化代谢增强了抗氧化能力,该结果也表明呼吸跃变和非呼吸跃变果实可能共享由褪黑激素介导的抗氧化反应。外源 MT 处理能通过增加 SOD 和 CAT 活性,降低 H_2O_2、O_2^-、电导率和 MDA 含量以及 PLD 和 LOX 活性,提高脯氨酸和 GABA 含量,从而抑制人心果发生冷害(Mirshekari et al., 2019)。但是 MT 在调控南果梨采后果实褐变中没有报道,关于其对膜脂代谢的调控机理也未见报道,为探索褪黑素对冷藏南果梨果皮褐变的影响及其调控作用机制,应用外源褪黑素处理南果梨,发现处理的果实在冷藏后常温货架期果皮褐变率和褐变指数明显降低,说明褪黑素处理能缓解冷藏南果梨果皮褐变的发生,我们又继续探索了其调控机制。通过第二章的细胞超微结果可以发现,褐变和非褐变南果梨果皮细胞内结构和组成有差异,因此,为了评估褪黑素对冷藏南果梨果皮褐变的影响,我们比较了褪黑素处理和对照果实果皮细胞超微结构。很明显,褪黑素处理能维持生物膜结构完整性,保护细胞器不受低温伤害,说明褪黑素处理可能对膜脂代谢有积极调控作用。细胞膜在冷胁迫下容易发生脂质过氧化,导致细胞膜氧化损伤,使细胞膜结构不稳定。在植物中,LOX 是负责膜脂质降解的关键酶(Yao et al., 2018),MDA 是膜脂过氧化最终产物(Lyons, 1973)。相对电导率是评估膜渗透性的有效参数,因此通常被认为是评估膜完整性的重要指标(Kong et al., 2018)。从前面结果可以看出,在冷藏后的常温货架期内,南果梨的相对电导率、MDA 含量和 LOX 活性明显增加,表明细胞膜结构不完整,膜脂质过氧化严重。在这里,外源 MT 处理可有效降低 MDA 含量和 LOX 活性,表明 MT 可以减少膜过氧化;此外,在 MT 处理过的果实中观察到较低的相对电导率和轮廓清晰、结构完好的细胞膜和细胞器,同时果皮褐变得到缓解,表明 MT 处理可以通过保护南果梨免受氧化损伤来维持细胞膜的完整性。从膜脂组分和膜脂代谢相关酶活性和基因表达情况来看，褪黑素处理果实 PC、PE 相对含量显著增加，PA 相对水平显著降低，不饱和脂肪酸相对含量增加，饱和脂肪酸相对含量降低，且脂肪酸不饱和度和不饱和指数明显低于对照果实，PLD 和 LOX 的活性及基因表达明显下降，此外，PuMYB21 和 PuMYB54 的转录水平也显著降低，表明褪黑素处理能通过抑制 PuMYB21 和 PuMYB54 的转录从而降低 PuPLDβ1 的转录，使 PLD 活性降低，在此过程中 LOX 活性和表达水平也显著降低，从而增加细胞膜脂的不饱和度，减少 PA 积累，维持了细胞膜结构完整性，也保护了酚酶和酚类植物的区室化分布，有效减少了酶促褐变的发生。这些结果证实了以前的研究，如 MT 处理降低了相对电导率和 MDA 含量，并赋予番茄耐冷性（Jannatizadeh et al., 2019）。类似的研究结果在黄瓜中也有报道，研究发现，褪黑素浸泡过的黄瓜果实相对电导率、MDA 和 ROS 含量较低，抗氧化酶活性如 SOD、CAT 和 APX 较高，叶绿体和线粒体的形变明显改善，脯氨酸含量显著提高，冷害指数显著低于对照果实（辛丹丹，2017）。在猕猴桃中，褪黑素处理能提高 SOD、CAT 和 APX 活性，并提高抗氧化物质如抗坏血酸和谷胱甘肽含量，从而增强猕猴桃果实抗氧化能力，降低相对电导率和 MDA 含量，抑制 O_2^- 和 H_2O_2 的积累，减少氧化损伤，达到延缓冷藏猕猴桃冷害的目的。与已有的研究结果类似，在本研究中，褪黑素处理的果实 SOD、CAT 和 APX 活性及相关基因表达明显提高，叶绿体形态清晰、结构完整，表明褪黑素处理能提高冷藏南果梨的抗氧化能力，降低细胞膜的氧化损伤。

第二节　多胺处理对采后果实品质的影响

多胺（Polyamines，PAs）是一种低原子量的小脂肪族胺，普遍存在于所有生物体内的含氮碱，并参与了广泛的生物过程，如植物生长发育（细胞分裂、分化、胚胎发生、坐果、果实成熟、开花和衰老）、生物和非生物胁迫的反应等（Barman et al., 2011；Chen et al., 2019；Galston et al., 1995；Kakkar et al., 2002）。在细胞 pH 值下，PAs 表现为阳离子，可以与 DNA、RNA、磷脂和某些蛋白质相互作用，还可以调节细胞膜的刚性和稳定性（Kaur-Sawhney et al., 1993；Ficker et al., 1994）。近年来，PAs 多用于果蔬的贮藏保鲜，包括桃、葡萄、芒果、西葫芦等（Qian et al., 2021；Mirdehghan et al., 2016；Zahedi et al., 2019；Palma et al., 2014）。在各种天然形式的 PAs 中，腐胺（PUT）、亚精胺（SPD）和精胺（SPM）是植物中鉴定的主要 PAs。在自然界中，PAs 通常以自由原子基的形式存在，此外，它们可以与酚酸等微小粒子（共轭结构）结合，然后又与不同的大分子（结合结构）结合。PAs 在植物体内的合成主要有两种方式：（1）脱羧酶催化精氨酸生成胍丁胺，最终形成腐胺（Putrescine，Put）；（2）鸟氨酸脱羧酶（Ornithine Decarboxylase，ODC）催化鸟氨酸生成 Put。精脒合成酶（Spermidine Synthase，SPDS）催化 Put 生成 Spd，因此 ODC 和 SPDS 在 PAs 的代谢中发挥重要作用。采后 PAs 处理导致的果实成熟延迟和保质期延长已在各种呼吸跃变和非呼吸跃变果实中被广泛报道（Singh et al., 2020；Khosroshahi et al., 2017）。Li et al.（2021）发现，用 PA 处理能通过保护线粒体结构和功能延缓冷藏南果梨褐变现象，PA 处理后 PA 含量及 PA 合成代谢酶活性及表达水平显著增加，ROS 水平降低；此外，PA 处理保存了果皮线粒体结构，具体表现为 PA 处理降低了线粒体膜通透性转换孔开放和线粒体膜通透性增加的速率和幅度。研究证实，外源应用 PAs 可通过增强膜稳定性、抗氧化能力和降低脂质过氧化、呼吸、乙烯生产和细胞壁降解酶的活性来延缓衰老和相关的代谢过程（Sharma et al., 2019）。为了延长青椒的采后货架期，在 4 ℃±1 ℃条件下，对青椒进行了外源施用亚精胺和腐胺（SPD-PUT，10 μmol/L+10 μmol/L、20 μmol/L+20 μmol/L、30 μmol/L+30 μmol/L）40 d 的试验。在整个贮藏期间，所有处理和未处理果实的可滴定酸、蛋白质含量、CAT 和过氧化物酶活性、叶绿素和辣椒素含量均逐渐降低。脯氨酸含量和抗氧化剂 1,1-二苯基-2-三硝基苯肼自由基（Diphenyl Picryl hydrazinyl Radical，DPPH）清除活性随着贮藏时间的延长而不断增加。根据所有形态和理化性状的反应评估出，在 3 个处理组合中，20 μmol/L SPD 和 20 μmol/L PUT 被认为是最佳的 SPD-PUT 组合，使青椒品质特性、酶活性和抗氧化活性在贮藏期内得到了保持（Patel et al., 2018）。黑李（*Syzygium cumini* L. Skeels）是一种热带和亚热带水果，具有多种药用价值，特别是在治疗糖尿病方面。其为非呼吸跃变特性果实，果实在成熟期采收时由于其薄皮和快速软化的特点使采后果实极易受到水分的损失和微生物的攻击，从而大幅降低其货架期。通常由于其高易腐性，黑李在自然条件下贮藏不会超过 2~3 d。为探究多胺处理对黑李果实品质和耐贮性的影响，在贮藏前用 0.5 mmol/L 和 1.0 mmol/L 浓度的 PUT 和 SPM 分别处理成熟果实，对照果实用蒸馏水处理，在 27 ℃±3 ℃，85%±5% RH 环境条件下贮藏 6 d，发现与对照组相比，1.0 mmol/L 的 PUT 处理最有效地减少了约 1.5 倍的重量损失和腐败。处理的果实还保留了比对照更高的花色苷和最小的脂质过氧化。在多胺处理下，总酚、类黄酮、抗坏血酸和抗氧化能力等生物活性物质的损失也最小。可溶性固形物含量在 1.0 mmol/L PUT 处理的果实中保留较高，但多胺处理不影响酸含量，该研究结果可为延长黑李果实的耐贮性、远距离销售和减少采后损失提供有益的参考（Mishra et al.，2022）。草莓属于小浆果水果，是天然抗氧化剂的良好来源。近几十年来，许多研究致力于延长草莓的货架期，并在采后条件下保持其营养价值。通过 Spm 和亚精胺 Spd（0 mmol/L、1.0 mmol/L 和 1.5 mmol/L）处理草莓，探究多胺处理对草莓果实采后 3 d、6 d、12 d 品质的影响。与对照相比，Spm 和 Spd 处理降低了果实在贮藏期间的失重率、腐烂率、可溶性固形物含量、果汁 pH 值和口感指数。增加了可滴定酸和维生素 C 含量、组织硬度、酚类化合物和抗氧化活性。因此 Spm 和 Spd 主要是通过通过提高果实的抗氧化活性和减少果实组织的破坏来防止果实的衰老和生物活性物质的损失。在所研究的处理中，喷施 1.5 mmol/L 的 Spm 和 Spd 是提高草莓果实贮藏寿命和品质的最佳处理方式，因此使用 Spm 和 Spd 可以作为改善草莓采后品质和货架期的一种有前途的方法，对草莓水果产业的发展具有显著的经济效益和环境效益（Jalali et al.，2023）。为研究亚精胺和精胺处理对甜椒果实品质、生物活性物质、抗氧化酶活性、货架期和衰老相关物质的影响，采用亚精胺和精胺（0.5mmol/L、1.0mmol/L 和 1.5 mmol/L）浸渍和壳聚糖（1%）涂膜处理辣椒果实。在 10 ℃、相对湿度 80%~90%条件下贮藏 56 d。结果表明，与其他处理相比，精胺 1.5 mmol/L 和壳聚糖 1%协同处理对减少贮藏期间的失重率、电解质渗漏、保持果实品质和增强抗氧化活性最为有效。与其他组相比，精胺 1.5 mmol/L 和壳聚糖 1%协同处理的甜椒果实在贮藏末期失重率降低了 47.6%，电解质渗漏率降低了 62.6%，MDA 含量降低了 28.3%，总酚含量降低了 68.8%，总类胡萝卜素含量降低了 21.9%，抗坏血酸含量降低了 15%，总叶绿素含量降低了 30.4%，CAT 活性降低了 34.3%。结果表明，多胺，特别是 1.5 mmol/L 精胺与壳聚糖协同处理对甜椒的货架期有潜在的有益影响，与其他组相比，甜椒的货架期延长了约 28 d，并显著地保持了生物活性物质、抗氧化酶和延缓衰老的相关指标（Sharma et al.，2022）。在压力渗透或浸泡下外源施加腐胺或亚精胺可以诱导石榴对低温的适应性，并通过增加内源腐胺和亚精胺的水平来保护果实免受低温冷害的影响（Pareek et al.，2015）。

Put 对果实细胞壁代谢酶的活性有一定的影响，在维持果实硬度和延长货架期上起重要作用，近年来被广泛应用于果蔬采后软化的研究中。Put 处理显著降低了辣椒 Cx 和 PME 的活性（Ghosh et al.，2021）。Put 处理有效维持了梨的硬度，抑制了淀粉和可滴定酸的降解以及 PE 和 Cx 的活性（Singh et al.，2020）。在芒果中，Put 处理抑制了果实成熟过程中乙烯的产生和细胞壁代谢酶活性，提高了抗氧化酶活性，从而维持了较好的果实品质（Razzaq et al., 2014）。Song et al.（2022）从果实细胞壁代谢的角度探讨了 Put 处理在蓝莓软化过程中的可能作用机制，用 1 mmol/L Put 溶液浸泡蓝莓 10 min，自然晾干后在 20 ℃放置 10 d。结果表明 Put 处理能有效的延缓蓝莓果实软化，与对照相比，Put 处理的蓝莓果实表现出更高的硬度和原果胶含量。此外，Put 处理显著抑制了细胞壁代谢酶的活性和表达水平，如 PG、β-Gal 和葡萄糖苷酶（β-glucosidase，β-Glu）。此外 PUT 处理还促进了 Put 合成基因 *VcODC* 的表达，抑制了 Put 代谢基因 *VcSPDS* 的表达。进一步研究表明，腐胺处理主要通过抑制细胞壁代谢关键基因 *Vcpg*1 的表达，延缓了采后蓝莓果实的软化。以上研究均表明外源 Put 处理可通过维持抗氧化化合物、减少脂质过氧化、抑制细胞壁降解、维持果实硬度等方式，有效地保持果实在采后贮藏期间的品质特性。

第三节　γ-氨基丁酸处理对采后果实品质的影响

γ-氨基丁酸（γ-aminobutyric acid，GABA）是一种四碳非蛋白质氨基酸，被认为是影响不同园艺作物采前和采后生理的不同方面的主要内源信号分子之一（Aghdam et al., 2016），特别是在生物和非生物胁迫条件下，在植物生长发育、逆境胁迫应答、细胞渗透调节、细胞供氮、清除自由基等方面发挥重要作用（Li et al., 2018；Aghdam et al., 2016；Shelp et al., 2017）。此外，GABA 被发现介导采后生理过程，如衰老、耐冷性和在冷藏期间保持园艺产品的质量（Aghdam et al., 2015；Aghdam et al., 2019a。Shelp et al., 2017；Niazi et al., 2021）。GABA 在 2009 年被我国确立为新资源食品，GABA 在发酵食品中的含量为较高，Takeshima 等（Takeshima et al., 2014）对 GABA 做了毒性的研究评估，结果显示 GABA 十分安全，能够作为一种功能性食品。作为一种新型食品功能性因子已被批准在食品中使用，而且美国环境保护署也已经确认这种物质在植物或果实上使用时对环境和人体健康无任何毒副作用（余辰，2014）。

GABA 通过 GABA 支路代谢，该支路由 3 种酶组成：谷氨酸脱羧酶（Glutamate Decarboxylase，GAD）、GABA 转氨酶和琥珀酸半醛脱氢酶，其中 GAD 是关键酶（Kinnersley et al., 2000）。在植物中，细胞内的 GABA 水平通常较低，但其可以在响应干旱、盐和低温胁迫时大量和迅速地积累，并参与对这些胁迫的防御过程（Kinnersley et al., 2000）。GABA 在脂质过氧化过程中作为

MDA 形成的抑制剂（Deng et al.，2010），维持膜完整性（Aghdam et al.，2015），增加抗氧化代谢和 GABA 分流途径，并有助于维持渗透调节（Aghdam et al.，2016；Li et al.，2016；Malekzadeh et al.，2017）。

GABA 处理园艺产品可通过诱导酶促和非酶促抗氧化反应、提高内源 GABA 含量、抑制 PG 和 PME 及多酚氧化酶活性、激活苯丙氨酸解氨酶活性等作用机制维持果实品质（Aghdam et al.，2016；Niazi et al.，2021；Wang et al.，2014；Yang et al.，2011）。最近，Song et al.（2010）报道了外源 GABA 可以缓解铝和质子胁迫对大麦幼苗造成的氧化损伤。然而，关于外源 GABA 处理作为影响采后园艺产品贮藏寿命和品质的因素的影响的信息很少。为了研究外源 GABA 对桃果实冷害的影响，新鲜采摘的桃子在 20 ℃的条件下用 1 mmol/L、5 mmol/L 或 10 mmol/L GABA 处理 10 min，然后在 1 ℃条件下贮藏 5 周。结果表明，所有 GABA 处理均能减轻桃果实冷害，其中 5 mmol/L 效果最好。GABA 处理可显著增加内源 GABA 和脯氨酸的积累，这与谷氨酸脱羧酶、Δ1-吡咯啉-5-羧酸合成酶和鸟氨酸 δ-氨基转移酶活性的增加以及脯氨酸脱氢酶活性的降低有关。表明 GABA 处理可能是减轻冷藏桃果实冷害的一种有效技术，而 GABA 对冷害的减轻可能是由于诱导内源 GABA 和脯氨酸的积累。这些数据是外源 GABA 诱导采后园艺产品耐冷性的首次证据（Shang et al.，2011）。为研究 GABA 对西葫芦果实采后冷藏品质、及其在 GABA 分流和其他相关代谢途径中的影响，用 1 mmol/L GABA 处理西葫芦。施用 GABA 提高了西葫芦果实在 4 ℃贮藏条件下的品质、降低了冷害指数、失重率，细胞死亡率和电解质外渗率。在所有分析的时间里，处理果实中的 GABA 含量显著高于对照果实。在贮藏期结束时，GABA 处理的果实具有较高的脯氨酸和腐胺含量，并且这种多胺的分解代谢不受外源 GABA 的影响。此外，GABA 处理还通过增加 GABA 转氨酶和 GAD 的活性来诱导 GABA 支路。GABA 处理的果实比未处理的果实含有更高水平的延胡索酸和苹果酸，以及更高的 ATP 和 NADH 含量。这些结果表明 GABA 支路可参与代谢物的提供以及能量的产生，降低冷害损伤，从而帮助果实长期应对冷胁迫（Palma et al.，2019）。外源应用 GABA 可降低梨中 ROS 积累，并降低膜脂质过氧化反应（Li et al.，2019）。为研究 GABA 处理对黑加仑葡萄采后品质、抗冷性和腐烂的影响，用不同浓度（0 mmol/L、20 mmol/L 和 40 mmol/L）GABA 处理黑加仑葡萄，与对照相比，40 mmol/L GABA 处理的果实失重率（35%）、果梗褐变（30%）和腐烂率（63%）较低。GABA 处理的葡萄中可诱导脱落酸产生从而导致较低的膜电解质渗漏率（13%）。此外，在 60 d 结束时，与对照组相比，40 mmol/L GABA

处理可产生更高的抗氧化酶活性，如 SOD（50%），CAT（35%），愈创木酚过氧化物酶（65%）和 APX（47%），以及更低的 MDA（21%）。40 mmol/L GABA 处理的葡萄由于多酚氧化酶活性降低，使得酚类物质（酚酸类、二苯乙烯类、黄酮类、花青素类）增加，抗氧化能力增强。因此，推荐 GABA 处理可用于保持葡萄采后贮藏期间的内在品质，减少真菌腐烂和冷害（Asgarian et al.，2022）。柑橘在我国占有重要的经济地位，因风味独特、富含营养物质而广受欢迎。靖安椪柑是江西省名特优柑橘品种，被称为"远东橘王"，具有果形美、色泽艳、汁多渣少、风味香甜、耐贮藏、成熟晚等特点，已成为靖安地区果农经济收入的重要来源（高阳等，2018）。然而，靖安椪柑生产具有很强的季节性，果实成熟期集中，采后贮藏存在果实糖和有机酸的减少、浮皮以及组织细胞的氧化损伤等问题，从而造成品质劣变，当地果农造成较大的经济损失（陈小红，2015；Gao et al.，2018；梁芳菲等，2018）。采后贮藏保鲜是柑橘生产中重要环节之一，为探究 GABA 处理对采后靖安椪柑果实品质和保鲜效果的影响，以靖安椪柑果实为试验材料，采用 0.5 mmol/L GABA 浸果处理后置室温（20 ℃±1 ℃）贮藏，测定果实失重率、色差指数、TSS、TA、可溶性糖和有机酸含量等品质指标，CAT、SOD、POD、APX 等抗氧化酶活性以及总酚、类黄酮、H_2O_2 含量和羟自由基清除率的变化。结果显示外源 GABA 处理能显著降低果实失重率，增加果皮亮度，维持果实 TSS、TA、总糖、可溶性糖和柠檬酸含量在较高水平，从而维持果实品质；同时，GABA 处理还显著提高椪柑果实贮藏中期（30~40 d）CAT、SOD 和 POD 活性以及贮藏前期（10 d）和后期（60 d）APX 活性，促进总酚和类黄酮含量积累，显著提高靖安椪柑果实羟自由基清除率，降低果实 H_2O_2 积累。说明 GABA 处理可能通过调节糖酸代谢、增强抗氧化酶活性和提高抗氧化物质含量以维持椪柑果实采后贮藏品质和保鲜效果（陈秀等，2022）。GABA 还影响果实采后有机酸代谢从而调控果实贮藏品质，苹果经 10 mmol/L GABA 处理可显著延缓可滴定酸和苹果酸的损失，并增加琥珀酸和草酸的含量（Han et al.，2018）；外源 GABA 处理能够抑制夏橙和纽荷尔脐橙果实贮藏期柠檬酸降解相关基因表达，有效减缓果实中柠檬酸含量下降，改善果实的贮藏品质（Sheng et al.，2017）。这些研究表明，GABA 有助于保持水果和蔬菜的采后品质。

第四节　L-谷氨酸处理对采后果实品质的影响

L-谷氨酸是一种广泛存在于植物、动物和微生物体内的氨基酸，通过分解代谢参与TCA。此外，它在氨基酸代谢中作为GABA合成的前体，并被整合到蛋白质中。L-谷氨酸和GABA在包括防御反应、信号通路和生殖功能在内的各种生理过程中都发挥着至关重要的作用（de Bie et al.，2023）。苹果（*Malus domestica* Borkh.）是世界上营养丰富的水果之一，水果中富含糖、酸、维生素、膳食纤维、类胡萝卜素和酚类化合物，其中一些对人类健康有特殊的功能。'秋金'苹果是一种主要种植在中国辽宁省的杂交品种，它因美丽的颜色、强烈的硬度和甜味而受到消费者的喜爱，然而，由于呼吸、软化、真菌感染和蒸发，采后苹果的质量下降（Huang et al.，2022）。研究表明，L-谷氨酸浸泡保持了'秋金'苹果具有较高的L^*、a^*和b^*值，果肉硬度、可滴定酸、可溶性固形物、可溶性糖、还原糖和抗坏血酸含量。L-谷氨酸处理还降低了果实的质量损失、呼吸速率和乙烯释放量，提高了SS裂解、AI和NI活性，降低了山梨醇脱氢酶、蔗糖磷酸合成酶、SS和山梨醇氧化酶活性。此外，L-谷氨酸抑制了叶黄素、β-胡萝卜素和番茄红素的积累，下调了八氢番茄红素合成酶、番茄红素β-环化酶、ζ-胡萝卜素去饱和酶、八氢番茄红素去饱和酶、类胡萝卜素异构酶、ζ-胡萝卜素异构酶和类胡萝卜素裂解双加氧酶基因的表达，但上调了9-乙酰氧基类胡萝卜素双加氧酶基因的表达。表明采后L-谷氨酸浸渍处理可以通过调节山梨醇、蔗糖和类胡萝卜素代谢关键酶活性和基因表达来保持苹果品质（Li et al.，2023）。L-谷氨酸还能促进氨基酸代谢和防御相关蛋白的积累，从而抑制梨的青霉病（Jin et al.，2019）。此外，L-谷氨酸已被证明可以诱导番茄果实对链格孢菌的抗性（Yang et al.，2017）。L-谷氨酸也可以通过调节氨基酸代谢来抑制鲜切马铃薯的褐变（Song et al.，2023）。'早酥'梨是中国西北地区的早熟品种，表面绿色光滑，质地清爽，风味丰富，营养丰富。然而，由于其采收期在夏季，采收时果实表现出强烈的呼吸作用，可导致果肉软化和果皮黄化，从而损害其原有的商业特性。这极大地缩短了货架期，给种植户和销售商造成了巨大的经济损失（Zhang et al.，2019）。为评估L-谷氨酸处理对'早酥'梨品质的影响，并阐明其潜在机制，用L-谷氨酸浸泡处理'早酥'梨果实，发现L-谷氨酸处理降低了乙烯释放量、呼吸强度、失重率、亮度（L^*）、红度（a^*）、黄度（b^*）和总色差（ΔE）；提高了抗坏血酸、可溶性固形物和可溶性糖含量；保持了梨的叶绿素含量和果肉硬度。L-谷氨酸还抑制NI和AI活性，同时增强蔗糖磷酸合成酶和SS活性，促进蔗糖积累。L-谷氨酸抑制*PbSGR1*、*PbSGR2*、*PbCHL*、*PbPPH*、*PbRCCR*和*PbNYC*的转录，导致叶绿素降解减慢。L-谷氨酸同时抑制PG、PME、纤维素酶和β-葡萄糖苷酶的转录水平和酶活性。它抑制了多聚半乳糖醛酸反式消除酶和果胶甲基反式消除酶的活性，并抑制了 *PbPL* 和 *Pbβ-gal* 的转录水平。此外，L-谷氨酸促进了精氨酸脱羧酶、鸟氨酸脱羧酶、S-腺苷甲硫氨酸脱羧酶、谷氨酸脱羧酶、γ-氨基丁酸转氨酶、谷氨酰胺合成酶的基因转录和酶活性以及 *PbSPDS* 的转录。L-谷氨酸还导致 *PbPAO*、*PbDAO*、*PbSSADH*、*PbGDH* 和 *PbGOGAT* 转录水平下调，同时提高梨中 γ-氨基丁酸、谷氨酸和丙酮酸含量。这些结果表明，L-谷氨酸浸泡处理可以通过调节蔗糖、叶绿素、细胞壁和多胺代谢的关键酶活性和基因转录来有效维持'早酥'梨的贮藏品质，L-谷氨酸浸泡在保持'早酥'梨品质的同时，表现出延缓果实褪黄和软化的能力（Jin et al., 2024），这些研究表明 L-谷氨酸有助于保持水果的采后品质。

第五节 赤霉素处理对采后果实品质的影响

赤霉素（Gibberellins，GAs）是生物体内的一类四环二萜类化合物，是公认的无毒害、无残留的植物生长调节剂，其用量少，成本低，能被分解（Teszlák et al., 2005）。至今已发现 100 多种，合并总称为赤霉素类（Kawaide et al., 2006）。大量研究表明，GA_3 具有很强的生物活性，对果蔬贮藏品质改善和采后生理变化有一定的作用（项丽霞，2006）。郑柯斌等（2020）发现 GA_3 可以提高蓝莓的可溶性固形物含量，改善果实口感。曹建康等（2008）发现，GA_3 能有效地降低鸭梨果实乙烯释放量，保持果实较高的硬度。袁莉等（2011）研究发现 GA_3 处理可有效降低枸杞果实腐烂率。采用 GA_3 处理采后莲藕可以延缓莲藕表皮 CAT 活性下降，同时显著抑制了过氧化物酶和多酚氧化酶活性，延缓莲藕的衰老（项丽霞，2006）。周传悦等（2023）发现，50mg/L GA_3 浸泡处理可有效延缓香蕉果实褐变，保持色泽和硬度，抑制 TSS 和抗坏血酸含量降低，降低 PPO 和 POD 酶活性，提高 CAT 和 APX 活性，控制了电导率和 MDA 含量的增加，有效地维持了细胞膜的结构和功能，说明赤霉素处理可延缓香蕉衰老进程，保持果实品质，延长贮藏期。赤霉素处理对提高甜樱桃的贮藏品质有较好的效果，赤霉素处理可以抑制 POD、CAT 和 PPO 活性的下降，并减少了 MDA 含量，使甜樱桃能够保持较高的鲜食品质水平（李夫庆等，2009）。黄铭慧（2015）研究发现，芒果在几种不同浓度的 GA_3 处理下均可以在一定程度上减缓硬度下降和色泽变化，延缓 TSS、维生素 C 和 TA 含量变化，并保持芒果果实的贮藏品质，经研究发现这与 GA_3 处理提高了果实 POD、SOD、CAT 活性、抑制 MDA 含量的升高、延缓果皮细胞膜透性上升有重要关系。采后不同浓度 GA_3 处理'贵妃'和'台农'芒果，与对照组相比，较低浓度的 GA_3 处理（0 g/L、1.0 g/L）能够显著抑制芒果的乙烯释放量，促进两个品种果实内赤霉素的合成。同时，较低浓度的 GA_3 处理（0.5 g/L、1.0 g/L）能够显著抑制果实软化，减少 MDA 合成，抑制果皮转色。而较高浓度的处理（2.0 g/L、3.0 g/L）没有显著的抑制软化和果皮转色的作用（李斯宇，2019）。以上研究结果说明赤霉素对采后果蔬品质维持具有正向调控作用。

第六节 糖处理对采后果实品质的影响

糖含量与果实的口感和品质密切相关，糖在植物体内具有重要的作用，是结构和能量贮存物质，也是呼吸底物和许多生化过程的中间代谢物，糖可作为信号分子参与调控植物的诸多生理生化过程及植物对生物和非生物胁迫的应答（Ciereszko，2018；Lastdrager et al.，2014；Holland et al.，2002；Wang et al.，2013；Zhao et al.，2021）。研究发现，葡萄糖、果糖和蔗糖等可溶性糖可作为渗透压调节剂、低温保护剂、ROS 清除剂及信号分子等多重作用来缓解果实的低温胁迫（Vichaiya et al.，2022；李佩艳等，2022；李姚瑶，2020）。Wang et al.（2016）与张杼润等（2019）研究发现，较高的葡萄糖与果糖水平可缓解杏果实的冷害现象。葡萄糖不仅是植物体内主要的能量供应物质，同时也可作为结构物质为合成酚类及黄酮类等物质提供碳骨架（Couee et al.，2006；Huang et al.，2013）。研究表明，外源葡萄糖的施用可使园艺作物提质增产、缓解植物逆境胁迫、调控植物成熟衰老（Couee et al.，2006；Huang et al.，2013；Wei et al.，2011）。Wang et al.（2021）研究发现，外源葡萄糖处理可有效保持采后草莓的贮藏品质，并提高其抗氧化能力。近年来，外源糖处理在果蔬采后方面的应用越来越多，已有研究表明外源糖处理可以有效影响果实的品质，如蔗糖处理能延缓芥菜叶绿素和蔗糖含量的下降，维持类胡萝卜素和抗坏血酸的含量，提高芥菜的抗氧化能力，有效保持了芥菜的品质（Di et al.，2022）。田梦瑶等（2022）研究发现，蔗糖处理能诱导桃果实花色苷的积累，提升果皮的色泽。蔗糖、葡萄糖处理可有效维持青花菜能量水平及增强其抗氧化能力，延缓青花菜的衰老与黄化（董栓泉等，2016；汤月昌等，2014）。葡萄糖还能显著提高甘蓝和小白菜芽中总硫代葡萄糖苷、酚类物质和花青素的含量，有效提高其品质（Wei et al.，2011）。对西兰花的研究也表明，果糖和蔗糖处理均能有效延缓采后西兰花的衰老及程序性死亡，抑制其品质的劣变（汤月昌等，2015；Han et al.，2019）。刘雪峰等（2021）采用外源蔗糖、葡萄糖、果糖对'塔罗科'血橙进行处理，结果表明葡萄糖处理可以显著提高血橙果面的光亮度。雷竹为禾本科竹亚科刚竹属竹种，雷竹幼体即为雷竹笋（Jiang et al.，2006）。雷竹滋味鲜美、硬脆爽口、富含各种维生素及矿物质等优良食用和营养特性，深受市场欢迎，笋体娇嫩，但采割时形成的笋体根部大面积机械伤可导致明显的笋体失水现象，加快笋体呼吸强度和生理衰老速度，并刺激创面及笋体中心处木质素及纤维素的大量积累，从而使笋体出现质地生硬、粗糙少汁和表面褐变等典型木质化败坏症状，严重降低了雷竹笋食用品质和商品性（Chen et al.，2013）。周大祥等（2020）发现 20 mmol/L 果糖喷淋处理可有效抑制雷竹笋冷藏期间笋体硬度上升、褐变和少汁等木质化败坏症状的发生，降低笋体失重率，维持较高的出汁率和感官品质，从而有效延长雷竹笋冷藏周期。研究发现 20 mmol/L 果糖处理可通过调控雷竹笋蔗糖代谢途径的方式，竞争性抑制笋体木质素的合成，从调控蔗糖代谢的角度分析了外源果糖处理减轻雷竹笋采后木质化败坏症状的机理。杏（*Prunus armeniaca* L.）在新疆林果业中占有重要的地位（李丽花等，2017），其营养丰富，深受消费者的喜爱（Fan et al.，2017）。杏属于呼吸跃变型果实，易后熟软化腐烂，极大地限制了杏果实的储存、运输、加工和销售（李亚玲等，2020）。张昱等（2023）研究发现，外源葡萄糖处理能通过调控杏果实贮藏期间蔗糖代谢相关酶 SS、SPS、AI、NI 活性，提高果实体内的葡萄糖与果糖含量，同时降低贮藏中后期蔗糖的含量，抑制 MDA 的积累与细胞膜渗透率的上升；同时，适宜浓度（200 mmol/L）的外源葡萄糖处理可保持杏果实的硬度、色泽、TSS、TA 与 ASA 含量，显著缓解了杏果实的冷害发病情况，证实外源葡萄糖处理可通过调控杏果实蔗糖代谢增强果实的抗冷性。研究发现，用 300 mmol/L 的果糖溶液进行减压渗透处理（0.05 MPa、2 min），之后恢复常压浸泡 5 min，该处理能显著抑制细胞壁降解酶 PG、PME、β-Gal 和 CEL 的活性，维持 NSP、CSP、纤维素含量，抑制细胞壁的降解和细胞膜渗透率的上升。细胞超微结构观察发现，果糖处理可显著维持杏果实细胞器和膜系统的完整性，有效延缓杏果实硬度的下降（芦玉佳等，2023）。以上研究结果均表明糖处理是一种有效的果蔬采后贮藏保鲜方法，可通过调节糖代谢、改善渗透压及影响相关酶活来维持果蔬良好贮藏的品质。

第七节 水杨酸处理对采后果实品质的影响

水杨酸（Salicylic Acid，SA）学名为邻羟基苯甲酸，是肉桂酸的衍生物

（边菊芳等，2009），是植物体内产生的一种简单酚类化合物（Hayat et al.，2007），在高等植物中广泛存在（曹伍林等，2014），与植物抗机械伤害和抗病性密切相关的信号分子（Gu et al., 2022），最初从柳树皮中分离出来，最早是应用于医疗（于继洲等，2004）。该物质能提高植物体内的各种防卫反应、影响植物多种代谢过程，如对病原菌和乙烯的产生具有直接抑制作用，还可降低呼吸速率，延缓植物衰老（Muzammil et al., 2015）。研究表明，外源施用 SA 处理对果蔬的发育、衰老、抗病性、乙烯合成等诸多生理过程有很大影响，可有效降低机械损伤对果蔬的伤害（Lu et al., 2011）。

外源 SA 处理对一些果蔬如猕猴桃（Huang et al., 2017）、番茄（Kant et al., 2016）、杏（郭科燕等，2012；马玄等，2015）、脐橙（陆云梅等，2011）、桃（蔡冲等，2004；韩涛等，2000）等具有良好的保鲜效果。外源 SA 或其衍生物乙酰水杨酸和水杨酸甲酯可以通过降低乙烯产生与呼吸速率，激活抗氧化系统和调节细胞壁的代谢成分，维持果实糖、酸和芳香类挥发性物质含量，降低果实的腐烂率、保持果实硬度、延缓果实衰老，维持果实的品质从而延长货架期（Asghari et al., 2010；李双芳等，2020；Zhou et al., 2018；Chen et al., 2023）。香椿（*Toona sinensis*），也叫椿芽，有着'树上青菜'的美称（尹雪华等，2017）。香椿芽中蛋白质、碳水化合物含量较高，富含多种维生素、矿物质及重要的生物活性成分如多酚、黄酮等（李双芳等，2020）。其不仅具有独特的清香，而且具有多种食疗功效（Sun et al., 2016）。每年的 3 月下旬至 5 月上旬是香椿采摘的季节，但由于新鲜椿芽水分含量大、生理代谢旺盛、呼吸强度大，容易出现失水萎蔫、褐变、脱叶、腐烂等现象，影响产品品质，对香椿延长供应带来较大困难（李双芳等，2020）。因此，对香椿采后保鲜处理至关重要。采用不同浓度（10 μmol/L、50 μmol/L、100 μmol/L）SA 浸渍处理香椿芽，置于（4±1）℃条件贮藏，对贮藏期间香椿芽的感官品质、营养指标和有害物质生成量变化进行监测，发现使用不同浓度 SA 处理均可延缓香椿品质下降，其中 50 μmol/Lol/L SA 处理效果最好，在贮藏后期仍表现出较好的感官品质，且贮藏期间花青素、维生素 C 及总黄酮含量均保持在较高的水平，此外还能有效地抑制采后贮藏期间亚硝酸盐含量的上升。说明水杨酸在降低香椿芽呼吸速率、保持香椿品质、延缓香椿衰老方面具有一定功效，且并不是水杨酸浓度越高保鲜效果越好，可能是外源水杨酸浓度过高会破坏细胞膜的完整性，加速水分的丧失。SA 安全环保，具有潜在的应用价值，为香椿的贮藏保鲜提供了重要参考（李双芳等，2020）。甜樱桃（*Prunus avium* L.）属蔷薇科李属樱亚属植物，又称西洋樱桃、大樱桃，是我国北方落叶果树中果实成熟最早的树种（谭维军等，2004）。其果实色、形、味俱佳，且含有丰富的酚类化合物和黄酮类化合物等多种营养物质，具有抗氧化性能，对预防心血管疾病、癌症、肿瘤及其他与氧化应激相关的疾病效果较好，素有"果中珍品"和"春果第一枝"的美称（刘淑清，2011；Aktaruzzaman et al.，2017）。甜樱桃于夏季成熟，温度较高，果实皮薄较脆弱，采后易出现果实软化、褐变、腐烂变质等现象，造成大量损失（杨艳芬，2009；施俊凤等，2009）。采后 2 mmol/L SA 处理可有效地抑制其果实腐烂的发生，延缓贮藏后期可溶性糖和维生素 C 含量的下降，延缓果实软化；还可抑制花青素含量的增加，抑制果实亮度的增加，促进果实变红变黄。但对延缓可溶性固形物和可滴定酸含量的下降没有明显的作用。此外 SA 处理可以抑制甜樱桃在贮藏过程中 H_2O_2 和 MDA 含量的升高，增强 SOD 活性，延缓贮藏中后期 CAT 活性的下降，但对贮藏过程中 POD 活性没有明显的作用（马艳艳，2018）。青椒（*Capasicum anmuum* L.）是茄科植物，原产于南美洲热带地区，于 20 世纪 70 年代传入我国（冯春婷等，2019）。青椒中富含维生素，具有促进人体肠胃蠕动、提高食欲等功效，深受人们的喜爱（马丽丽等，2021）。但采后青椒在运输过程中易受到机械损伤，组织结构遭到破坏，易遭受微生物侵染，引起果实生理代谢紊乱，加速果实的劣变，严重影响其商品价值（Wang et al.，2018）。桑兆泽等（2022）研究发现，5 mmol/L SA 处理能显著延缓青椒转红、软化及腐烂等衰败特性，提高 POD、CAT 和 APX 的活性，降低 MDA 的积累，并减缓了维生素 C 的降解，说明 SA 处理可以有效延缓青椒果实机械损伤后品质的劣变与衰老进程，延长青椒果实的货架期。杏属蔷薇科李属，是典型的呼吸跃变型果实，杏果实采收后成熟衰老非常迅速，腐烂损失严重，在采后运输和贮藏的过程中容易造成大量的经济损失（刑军等，2005）。研究发现采后 0.01g/L SA 处理能够显著降低杏果实在低温贮藏期间的冷害发病率、冷害指数、细胞膜透性和 MDA 含量，可有效保持杏果实的硬度与出汁率。表明外源 SA 处理可以提高杏果实的抗冷性（袁洁等，2013）。为研究外源 SA 处理对蓝莓果实采后生理及贮藏品质的影响，分别用 0.5 mmol/L、1.0 mmol/L 和 1.5 mmol/L 的 SA 溶液处理"夏普蓝"蓝莓果实，测定其感官品质、硬度、可溶性蛋白质、MDA、总酚及相关抗氧化酶活性，发现外源 SA 处理能通过提高总酚和可溶性蛋白含量，提升 SOD 和 POD 活性，降低 MDA 含量，有效延缓蓝莓果实衰老，显著抑制果实采后生理代谢，保持其低温贮藏品质（张瑜瑜，陈泽斌等，2022）。冬枣（*Ziziphus Jujuba* Mill. cv. Dongzao）是鼠李科中的枣属植物，也称雁来红、苹果枣、冰糖枣等，在我国主要分布在新疆、山东、河北等地。冬枣皮薄多汁，口感甘甜清脆，可食率高达96.9%（Zhao et al.，2021）。冬枣具有较高的营养价值，富含K、Na、Fe、Cu等矿物元素，以及氨基酸、维生素、多酚、类黄酮、功能性多糖、环磷酸腺苷等生物活性化合物。冬枣中维生素C含量是草莓的70倍，梨的100倍，因此荣获"百果王""活维生素丸"和"天下奇果"等美誉（韩齐齐等，2021）。冬枣具有调节脾胃、解毒保肝、增强免疫功能和心肌收缩力、降血压、降胆固醇、预防心血管疾病、癌症、美容养颜及抗氧化等功效，作为一种保健食品深受消费者青睐（李志欣等，2019；张婷婷，2020）。然而，冬枣皮薄肉脆、含水量较高，采后贮藏不当会导致冬枣快速成熟和衰老，严重影响其营养价值和经济价值（Zozio et al.，2014）。桑月英（2022）研究发现，0.5 mmol/L SA处理可显著延缓冬枣果实硬度的下降并保持糖酸水平，还可以通过抑制呼吸强度，降低乙烯生成量，降低贮藏期间的失重率和腐烂率。外源SA处理可通过诱导冬枣采后抗氧化活性（如提升SOD、APX和GR酶活性，提高抗坏血酸、谷胱甘肽、总酚和类黄酮含量）以及苯丙烷代谢（如提高PAL、C4H、4CL酶活性），从而提高抗氧化能力。葛阳杨等（2014）研究发现，0.01%水杨酸处理使冬枣样品的硬度指标优于未处理的冬枣样品。SA还可抑制"丰水"梨果实硬度损失和细胞壁降解，维持可溶性糖和总酚含量，进而保持采后梨果实品质和延缓成熟（Zhang et al.，2023）。最佳浓度的SA处理也可以有效地减轻冷害，维持采后梨果实品质，减少果实腐烂和组织褐变的发生（Hassan et al.，2007；Adhikary et al.，2020；Sinha et al.，2022；Luo et al.，2022）。以上研究结果均表明SA是一种有效的果蔬采后贮藏保鲜剂，可通过提高抗氧化系统维持果蔬良好的品质。

第八节 抗坏血酸处理对采后果实品质的影响

抗坏血酸（Ascorbate Acid，AsA）也叫维生素C，是植物体内的一类小分子抗氧化剂物质，有着良好的抗氧化性，能直接清除ROS，在维持细胞膜结构、减少膜脂过氧化和延缓衰老等方面有重要作用（刘拥海等，2011；林植芳等，1985；赵会杰等，1992；Kunert et al.，1985）。大量研究表明，适宜浓度AsA处理可以延长水晶梨（田密霞等，2008）、杏（杜善保等，2007）、葡萄（王文举等，2010）、荔枝（莫亿伟等，2010）圣女果（刘锴栋等，2012）等园艺产品的贮藏期。胡萝卜没有生理休眠期（刘恩信等，1997），由于贮藏过程中自身养分的不断消耗，使其营养品质不断下降。杨娜等（2012）采用100 mg/L AsA处理'京红五寸'与'鞭杆红'品种的胡萝卜，在 0 ℃条件下贮藏后定期测定胡萝卜的品质指标，研究外源 AsA 处理对胡萝卜贮藏过程中品质的影响。发现外源 AsA 处理能保持胡萝卜蛋白质含量及 AsA 含量，延缓胡萝卜糠心现象的发生、并能延缓胡萝卜游离氨基酸含量的升高；外源 AsA 处理提高了胡萝卜 CAT 和 POD 的活性，抑制了胡萝卜贮藏过程中 H_2O_2 含量的增加，表明外源 AsA 处理可有效延缓胡萝卜在贮藏过程中的衰老进程，能延长胡萝卜的贮藏时间。王静（2012）发现外源 AsA 处理可延缓龙眼果实果皮褐变的发生，降低呼吸速率，延迟果皮转色，保持较高的好果率，还能保持较高的 SOD、CAT、APX 等 ROS 清除酶活性和 GSH、AsA、类黄酮等内源抗氧化物质含量，降低龙眼果皮超氧阴离子（$O_2^{-·}$）的产生速率和膜脂过氧化物 MDA 的积累，增强 ROS 清除能力，保护细胞膜结构完整性。王静（2015）还研究了外源 AsA 对陕西省周至县主栽猕猴桃品种'秦美'的保鲜效果。发现 5% 的 AsA 处理可以有效降低'秦美'猕猴桃果实的呼吸速率和乙烯释放速率来调控乙烯代谢和呼吸作用，延缓细胞膜透性的升高和硬度的下降，减少营养成分的消耗，减缓可溶性固形物、可溶性总糖含量升高，保持较高的果肉维生素 C 含量，减少果实的失重和腐烂，从而维持较高的猕猴桃果实好果率。香蕉是典型的呼吸跃变型热带水果，极易在采后发生褐变、软化等现象。杨菊等（2022）探究了 50 mmol/L AsA 处理对香蕉 ROS 代谢、抗氧化能力和贮藏品质的影响，发现 AsA 处理能有效提高贮藏期间香蕉 SOD、CAT、APX 抗氧化酶活性，降低 H_2O_2 含量，从而减缓香蕉衰老软化，延缓褐变发生。综上所述，AsA 作为一种植物内源的抗氧化剂，可通过提高抗氧化酶活性，增加抗氧化物质含量，提升采后果蔬的贮藏品质。

第九节　涂膜处理对采后果实品质的影响

将保鲜液涂抹在果实的表面为涂膜保鲜，通过涂膜保鲜可以阻隔外界环境中的有害影响，从而抑制果实表面微生物生长和果实的呼吸作用，实现对果蔬保鲜的目的。苹果保鲜过程中常用到的涂膜材料是壳聚糖、蛋白质、聚乙烯醇、多糖类蔗糖酯等（阿地拉·阿不都拉，2021）。采用涂膜贮藏保鲜在市场上的应用也较广，涂膜保鲜对于甜瓜的硬度、水分的保持、营养物质的消耗减少等效果比较好，但是只能作为短期销售的一种辅助方法，且涂膜的厚度会直接影响到果实的风味和品质（王兰菊等，2003）。采前麝香草酚微胶囊/魔芋葡甘露聚糖/低酰基结冷胶复合可食性涂膜处理减缓了蓝莓果实贮藏过程中感官品质、硬度、L^*、b^*、果形指数、咀嚼性和弹性的下降；抑制了质量损失率、腐烂指数、MDA、还原糖含量的增加；延缓了可滴定酸、总糖、果实色泽 a^*、C^* 值的变化，可将贮藏期从 28 d 延长至 42 d（黄彭等，2024）。将魔芋葡甘聚糖、壳聚糖、蓝莓叶多酚制膜，可降低蓝莓贮藏过程中叶绿素和维生素 C 的损失，能有效抑制果实 MDA 生成，延缓蓝莓衰老（张颖，2017）。以 1.0% 羧甲基壳聚糖和 0.5% 香茅草精油制备的复合涂膜材料可提高沃柑果皮中苯丙氨酸解氨酶、过氧化物酶、多酚氧化酶、CHI 等抗病相关酶的活性及总酚、类黄酮、花青素等抗性物质的含量，增强采后沃柑抗病能力，有效保持果实品质（杜倩洁等，2024）。采用 GABA-壳聚糖涂膜处理能显著抑制人参果的呼吸作用，降低 PPO 和 POD 酶活性，维持 TSS、TA、维生素 C 和总黄酮含量，提高感官品质，延缓果实软化及品质劣变进程（梁金甜等，2024）。茶多酚和壳聚糖复配能够延长草莓的保质期（冯文婕等，2016）。茶多酚与壳聚糖复配还可显著减缓葡萄果实在贮藏期的质量损失，降低果实中可溶性糖、维生素 C 和 TA 含量下降，保持果实品质（曹婷等，2016）。1% 和 2% 的茶多酚溶液与 1% 的海藻酸钠溶液复配涂膜可显著降低葡萄在贮藏期间的失重率和腐烂率，保持较高的硬度、TSS 和 TA 含量，抑制 POD 活性的升高，提升贮藏期间葡萄果实的品质（石飞等，2023）。李东等（2024）用 2% 氯化钙浸泡后结合聚乙烯醇涂膜复合处理采后脆红李果实，发现该涂膜处理可有效延缓脆红李的衰老变质。20 g/L 鱼皮明胶和 10 g/L 壳聚糖复合涂膜处理可降低西番莲果实的呼吸速率和乙烯释放速率，延缓果实细胞膜透性和 MDA 含量的升高，降低失重率，保持西番莲果实较好的色泽，维持果实 TSS、TA、维生素 C 和总糖等营养物质含量，表明该复合涂膜处理可改善采后西番莲果实的贮藏品质，增强其耐贮性（陈洪彬等，2022）。以上研究结果均证实复合涂膜处理可有效维持采后果蔬品质，是一种可行的生物保鲜方法。

第十节 其他处理对采后果实品质的影响

乙酰水杨酸是一种能明显诱导植物抗病性的小分子信号物质，可在植物体内转换为 SA（Cezar et al.，2015；Zhou et al.，2015；Zhu et al.，2016）。研究发现，2 mmol/L 乙酰水杨酸处理可有效抑制采后'富士'苹果果心和果肉组织病斑直径扩展、MDA 含量增加和 LOX 酶活性升高，降低 DHA、GSSG 的含量，提高果实中抗氧化剂 AsA 和 GSH 含量，增加果实中抗氧化相关酶 APX、GR、DHAR 和 MDAR 活性，增强果实的 ROS 清除能力，消除过量的 H_2O_2，降低膜脂过氧化程度（邓惠文等，2021）。外源 ABA 处理可通过提高桃果实贮藏期间 SOD、POD、APX、GR 活性和相关基因表达，提高果实的抗氧化能力；还可促进磷酸 SPS 和 SS 的活性以及其基因表达，抑制 AI 和 NI 的活性及基因表达，维持果实硬度、山梨醇、蔗糖和 TSS 含量，抑制呼吸速率和乙烯的释放，减轻桃果实的冷害症状；此外 ABA 处理对草莓保鲜也有一定的效果，主要表现为通过抑制细胞壁降解相关基因（*FaPME*、*FaPG*、*FaPLA*、*FaGal*、*FaXTH* 和 *FaAra*）的表达从而保持草莓超微结构的相对完整性；通过抑制乙烯释放和呼吸速率，降低草莓腐烂率和失重率（唐继兴，2023）。外源柚皮苷处理可显著延缓柑橘果实病害，降低 H_2O_2 含量，提高抗氧化能力，维持苯丙烷代谢途径中橙油素、紫铆黄酮、柚皮素和木犀草素类化合物的含量，提升果实品质（Zeng et al.，2022）。枇杷果实喷洒重组丝氨酸蛋白酶贮藏 25 d 后，表面未见明显腐烂，TSS 和 AsA 含量显著升高，抗性相关酶活性增加，致病菌被抑制，表明重组丝氨酸蛋白可通过提高抗氧化能力，提升枇杷果实的抗病能力，诱导启动防御反应，起到防腐作用，有效维持枇杷果实的贮藏品质和延长货架期，具有较好的保鲜效果（Yan et al.，2021）。暹罗芽孢杆菌是一种芽孢杆菌，暹罗芽孢杆菌浸泡处理芒果后于 25 ℃贮藏过程中，能够有效地抑制病斑形成，减少 MDA 积累，延缓硬度下降，保持较高的类黄酮含量，维持 POD、CAT、β-1,3-葡聚糖酶（GLU）、CHI、PAL 和查尔酮异构酶活性，表明暹罗芽孢杆菌可有效诱导芒果果实抗病相关基因表达，抑制病害的发生，维持芒果果实的贮藏品质（李佳怡等，2024）。ε-聚赖氨酸是一种天然、高效的生物防腐剂。ε-聚赖氨酸雾化处理采后蟠桃果实，可降低其腐烂率和失重率，维持较高的色泽、可滴定酸含量和维生素 C 含量，抑制果实呼吸强度和乙烯生成速率，扫描电镜显示，ε-聚赖氨酸处理可保持果实细胞结构完整性，提高苯丙烷代谢相关酶活性（包括 PAL、POD、4CL、C4H）和总酚、类黄酮、木质素含量，提升 DPPH 自由基清除率和总抗氧化能力，表明 ε-聚赖氨酸处理可通过提高果实的苯丙烷代谢和抗氧化能力，降低病原菌的侵染，提升蟠桃的抗病性（邵丽梅，2023）。ε-聚赖氨酸处理也能通过诱导苹果抗性相关的 PPO、POD、PAL 和 CHI 活性及其编码基因的上调表达，促进苹果分泌总酚、类黄酮和木质素等抗性相关物质，调节苹果的呼吸代谢、ROS 代谢和膜脂代谢增强苹果采后青霉病的抗性，控制青霉病的发生（窦勇，2023）。茉莉酸甲酯（Methyl Jasmonate，MeJA）是茉莉酸（Jasmonic Acid，JA）在茉莉酸羧甲基转移酶的作用下转化而来，是一种内生激素（Nuez-gómez et al.，2020）。作为一种天然的植物调节剂，MeJA 已被证实能有效提高采后果蔬的抗氧化性和抗病性，延缓果蔬衰老进程（Fza et al.，2019）。研究发现，MeJA 处理可显著抑制 AI、NI 和 SS 的酶活性，同时显著提高蔗糖磷酸合成酶的酶活性，进而维持较高的蔗糖含量和贮藏品质，延长货架期；不同浓度的 MeJA 均可有效维持采后荔枝的贮藏品质，其中 50 μmol/Lol/L 的 MeJA 效果最佳（殷菲胧等，2024）。用 10 μmol/L MeJA 熏蒸树莓果实 8 h，可维持果实硬度、TSS、TA、抗坏血酸及总酚含量（马大文等，2018）。猕猴桃经过 0.1 mol/L 茉莉酸甲酯熏蒸 24 h，可明显降低果实的失重率和腐烂率（盘柳依等，2019）。咖啡酸可通过调节脂肪酸代谢增强苹果果实的贮藏能力（Huang et al.，2022）。

参考文献

阿地拉·阿不都拉, 2021. 三种保鲜剂对苹果贮藏过程中品质的影响 [D]. 阿拉尔: 塔里木大学.

艾沙江·买买提, 张校立, 梅闯, 等, 2018. 库尔勒香梨果实可溶性糖积累及代谢相关酶活性变化 [J]. 新疆农业科学, 55 (4): 664-673.

白琳, 吕静祎, 孙琨, 等, 2021. 氯化钙处理对采后南果梨硬度及相关酶活的影响 [J]. 渤海大学学报 (自然科学版), 42 (3): 210-216.

边菊芳, 徐祥彬, 薛大伟, 等, 2009. 植物系统获得抗病性的水杨酸信号传递机制研究进展 [J]. 杭州师范大学学报 (自然科学版), 8 (3): 224-228.

卜庆状, 2012. 不同处理对冷藏南果梨常温后熟过程香气变化的影响 [D]. 沈阳: 沈阳农业大学.

蔡冲, 陈昆松, 贾惠娟, 等, 2004. 乙酰水杨酸对采后玉露桃果实成熟衰老进程和乙烯生物合成的影响 [J]. 果树学报, 21 (1): 1-4.

曹建康, 李庆鹏, 姜微波, 等, 2008. 赤霉素处理对鸭梨果实乙烯代谢和贮藏品质的影响 [J]. 中国农学通报, 24 (1): 81-84.

曹婷, 朱明, 丁莎莎, 等, 2016. 茶多酚复配剂对新美人指葡萄贮藏品质的影响 [J]. 江苏农业科学, 44 (5): 336-339.

曹伍林, 宋琦, 孟祥才, 2014. 外源水杨酸在园艺植物栽培中的应用前景 [J]. 北方园艺 (16): 191-193.

曹香梅, 2019. 桃果实酯类芳香物质的代谢与调控研究 [D]. 杭州: 浙江大学.

曹中权, 余璐璐, 徐飞, 2016. 没食子酸丙酯处理对薄皮甜瓜采后储藏保鲜的影响 [J]. 北方园艺 (5): 148-152.

常雪花, 王振菊, 李忠, 等, 2019. 一氧化氮熏蒸对冬枣采后贮藏品质的影响 [J]. 食品工业, 40 (3): 176-180.

陈海荣，申琳，欧阳丽喆，等，2007. 一氧化氮对采后油菜品质与活性氧代谢相关酶的影响［J］. 食品科学，28（7）：493-496.

陈洪彬，李书亮，蒋璇靓，等，2022. 鱼皮明胶-壳聚糖复合涂膜对'黄金'西番莲的保鲜效果［J］. 食品与发酵工业，48（18）：8.

陈金印，曾荣，李平，2003. 猕猴桃果实冷藏过程中生理生化变化［J］. 食品科学（2）：138-141.

陈敬鑫，张德梅，李永新，等，2021. 低氧贮藏对采后果实风味的影响研究进展［J］. 食品科学，42（13）：273-280.

陈昆松，郑金土，张上隆，等，1999. 乙烯与猕猴桃果实的后熟软化［J］. 浙江农业大学学报，25（3）：251-254.

陈雷，秦智伟，1999. 甜瓜采后生理和贮藏保鲜研究进展［J］. 北方园艺（6）：24-27.

陈丽娟，杨金初，王赵改，等，2015. 外源甜菜碱对香椿嫩芽采后品质的影响［J］. 食品与生物技术学报，34（12）：1315-1320.

陈留勇，孔秋莲，孟宪军，等，2003. 浸钙处理对黄桃后熟软化的影响［J］. 食品科技，28（7）：22-24.

陈琪琪，杨洋，郭丽红，等，2023. 果蔬低温贮藏的糖代谢转录调控研究进展［J］. 食品研究与开发，44（8）：207-212.

陈蓉，詹志鹏，李万飞，2011. 菠菜中绿色素的提取及其稳定性研究［J］. 饮料工业，14（5）：18-21.

陈熙，苏宇萌，郭家如，等，2024. 基于能量代谢探究不同温度下马铃薯贮藏品质变化规律［J］. 食品与发酵工业，50（6）：159-168.

陈小红，2015. 采收期、贮藏温度和热水处理对靖安椪柑果实贮藏保鲜效果的研究［D］. 南昌：江西农业大学.

陈秀，罗震宇，张亚男，等，2022. 外源GABA处理对采后靖安椪柑果实品质和保鲜效果的影响［J］. 果树学报，39（4）：652-661.

程南谱，王鑫，ASLAM M M，等，2024. 荔枝 *LcWRKY*47 基因的克隆、亚细胞定位、表达及延缓果实褐变的功能初探［J/OL］. 热带作物学报，45（10）：2025-2034.

程顺昌，任小林，饶景萍，等，2005. 一氧化氮（NO）对采后青椒某些生理生化特性与品质的影响［J］. 植物生理学通讯，41（3），322-324.

程顺昌，2013. 不同采收期南果梨采后褐变发生机理及调控技术研究［D］. 沈阳：沈阳农业大学.

寸丽芳, 2023. 外源钙参与调控骏枣果实裂果及其作用机制的研究 [D]. 阿拉尔: 塔里木大学.

单秀峰, 徐方旭, 2015. 低温贮藏对甜玉米采后生理品质的影响 [J]. 沈阳师范大学学报 (自然科学版), 33 (4): 507-510.

邓惠文, 王应强, 韩雍, 等, 2021. AsA 处理对苹果果实抗坏血酸—谷胱甘肽循环和膜脂过氧化水平的影响 [J]. 陇东学院学报, 32 (5): 92-97.

董萍, 2011. 南果梨香气成分分析及其在采后贮藏过程中的变化 [D]. 沈阳: 沈阳农业大学.

董栓泉, 熊茜, 王春幸, 等, 2016. 蔗糖在延缓青花菜黄化过程中维持其能量和抗氧化力 [J]. 园艺学报, 43 (9): 1825-1833.

窦勇. 2023. ε-聚赖氨酸对苹果采后青霉病的控制及其机制研究 [D]. 镇江: 江苏大学.

杜倩洁, 杨阔, 周慧琴, 等, 2024. 沃柑采用香茅草精油-羧甲基壳聚糖复合液涂膜保鲜的效果 [J]. 中国南方果树, 53 (3): 28-36.

杜善保, 邹养军, 2007. 外源抗坏血酸对杏果实采后衰老的影响 [J]. 陕西农业科学 (5): 37-39.

杜秀敏, 殷文璇, 赵彦修, 等, 2001. 植物中活性氧的产生及清除机制 [J]. 生物工程学报, 17 (2): 121-125.

杜艳民, 王文辉, 贾晓辉, 等, 2021. 前期低氧处理对梨虎皮病的防控及乙烯释放的影响 [J]. 园艺学报, 48 (1): 15-25.

范林林, 高丽朴, 王清, 等, 2015. 外源 NO 处理对豇豆采后生理特性的影响 [J]. 食品与发酵工业, 41 (10): 191-196.

范中奇, 邝健飞, 陆旺金, 等, 2015. 转录因子调控果实成熟和衰老机制研究进展 [J]. 园艺学报, 42 (9): 1649-1663.

冯春婷, 陶永清, 董成虎, 等, 2019. 不同复合保鲜剂处理对青椒采后贮藏品质的影响 [J]. 食品研究与开发, 40 (19): 95-99.

冯文婕, 阙斐, 陈岭, 等, 2016. 茶多酚-壳聚糖复合涂膜液对草莓保鲜效果的研究 [J]. 现代农业科技 (22): 260-262.

冯叙桥, 孙海娟, 何晓慧, 等, 2013. 1-MCP 应用于芒果贮藏保鲜的研究进展 [J]. 食品与生物技术学报, 32 (12): 1233-1243.

付润山, 姜妮娜, 饶景萍, 等, 2010. 赤霉素和萘乙酸对柿果实采后成熟软化生理指标的影响 [J]. 西北植物学报, 30 (6): 1204-1208.

高红豆, 胡文忠, 管玉格, 等, 2021. 采后果蔬呼吸代谢途径及其调控研究进展 [J]. 包装工程, 42 (15): 30-38.

高慧, 饶景萍, 王毕妮, 等, 2010. 冷害与油桃果实采后生理及贮藏品质的关系 [J]. 食品与发酵工业, 36 (9): 181-185.

高婧宇, 谢龙莉, 陈楠, 等, 2024. 低共熔溶剂在类胡萝卜素提取上的应用研究进展 [J]. 食品工业科技, 45 (1): 352-358.

高启明, 侯江涛, 李疆, 2005. 库尔勒香梨生产现状与研究进展. 中国农学通报, 21 (2): 233-236.

高阳, 阚超楠, 陈楚英, 等, 2018. '靖安椪柑'果实发育阶段柠檬酸含量变化及其代谢相关基因的表达分析 [J]. 果树学报, 35 (8): 936-946.

葛阳杨, 郑钢英, 金姣姣, 等, 2014. 冬枣的保鲜方法研究及电子鼻评价. 中国食品学报, 14 (12): 205-210.

关军锋, 1992. 新红星苹果采后膜质过氧化的变化及其调节因素 [J]. 河北果树 (3): 19-22.

郭光艳, 柏峰, 刘伟, 等, 2015. 转录因子对木质素生物合成调控的研究进展 [J]. 中国农业科学, 48 (7): 1277-1287.

郭慧静, 金新文, 张有成, 等, 2023. 冬枣产业现状及保鲜技术研究进展 [J]. 安徽农业科学 (23): 1-4, 8.

郭科燕, 左宝莉, 贾盼盼, 等, 2012. 水杨酸处理对杏果实贮藏品质的影响 [J]. 食品工业科技, 33 (15): 335-337.

郭芹, 王吉德, 李雪萍, 等, 2013. 一氧化氮处理对采后番木瓜果实乙烯生物合成的影响 [J]. 广东农业科学, 40 (3), 75-78.

郭芹, 吴斌, 王吉德, 等, 2011. NO 处理对番木瓜采后贮藏性的影响 [J]. 食品科学, 32 (4): 227-231.

郭艳华, 张玉敏, 李艾华, 等, 2015. 一种天然绿色素改性产物的光热稳定性研究 [J]. 江汉大学学报（自然科学版）, 43 (4): 303-307.

韩齐齐, 张娅妮, 冯荤荤, 等, 2021. 冬枣采后生理与气调贮藏关键技术研究 [J]. 食品与发酵工业, 47 (4): 33-39.

韩涛, 李丽萍, 2000. 水杨酸处理对桃贮藏期间活性氧代谢的影响 [J]. 北京农学院学报, 15 (4): 41-47.

韩絮舟, 吕静祎, 白琳, 等, 2020. 采后氯化钙处理对红树莓保鲜的影响 [J]. 食品工业科技, 41 (6): 233-238, 243.

郝利平，寇晓虹，1998. 梨果实采后果心褐变与细胞膜结构变化的关系［J］. 植物生理学通讯，36（6）：471-474.

何双，李高阳，蒋成，等，2019. 气调贮藏对黄桃风味和软化及褐变相关酶的影响［J］. 现代农业科技（7）：202-205，218.

何伟，2020. 果蔬气调保鲜技术及其在冷链物流中的应用研究进展［J］. 食品与机械，36（9）：228-232.

侯冬岩，回瑞华，杨梅，等，2005. 南果梨中总黄酮的光谱分析及抗氧化性能测定［J］. 食品科学，26（2）：193-196.

胡丽松，吴刚，郝朝运，等，2017. 菠萝蜜果实中糖分积累特征及相关代谢酶活性分析［J］. 果树学报，34（2）：224-230.

胡苗，2018. 采后褪黑素处理对'华优'猕猴桃果实冷害和成熟衰老的影响［D］. 杨凌：西北农林科技大学.

胡宇微，孙红男，木泰华，2023. 提高叶绿素稳定性方法的研究进展［J］. 食品科技，48（2）：49-55.

黄方，迟英俊，喻德跃，2012. 植物MADS-box基因研究进展［J］. 南京农业大学学报，35（5）：9-18.

黄方，唐杰，黄敏，等，2022. 低温结合气调包装对荔枝保鲜作用［J］. 食品工业，43（10）：51-55.

黄铭慧. 2015. 外源赤霉素处理对'贵妃'芒果贮藏品质、采后生理及催熟品质的影响［D］. 海口：海南大学.

黄彭，丁捷，刘春燕，等，2024. 可食性复合涂膜对蓝莓的保鲜效果综合评价［J］. 园艺学报，51（6）：1361-1376.

黄玉平，2016. 外源NO对草莓果实品质影响及诱导抗病性机理研究［D］. 南京：南京农业大学.

黄泽军，黄荣峰，黄大昉，2002. 植物转录因子功能分析方法［J］. 农业生物技术学报（3）：295-300.

及华，关军锋，窦世娟，等，2004. 赞皇大枣采后成熟衰老期生理生化变化［J］. 河北农业科学，8（2）：13-16.

及华，关军锋，冯云霄，等，2005. MAP在果蔬贮藏保鲜中的应用效果及其作用［J］. 保鲜与加工（1）：7-10.

纪淑娟，范婷，程顺昌，等，2012. 采收成熟度对南果梨货架期褐变的影响［J］. 食品工业科技，33（14）：334-338.

纪迎琳，蔺世姣，王盼盼，等，2022. 乙烯利、脱落酸和1-MCP对早酥和早金酥梨果实后熟的影响 [J]. 沈阳农业大学学报, 53 (5): 513-519.

贾海锋, 赵密珍, 王庆莲, 等, 2016. 生长素和脱落酸在草莓果实发育过程中的作用 [J]. 江苏农业科学, 44 (11): 173-176.

贾晓辉, 张鑫楠, 王东峰, 等, 2023. 不同气体组分对玉露香梨采后生理及品质的影响 [J]. 中国果树 (8): 17-23.

姜巍, 2015. 南果梨黑星病的危害现状及防治措施 [J]. 农业开发与装备 (2): 121-122.

焦彩凤, 林琼, 柴奕丰, 等, 2019. NO 处理对采后果实的保鲜作用及作用机制 [J]. 食品安全质量检测学报, 10 (2): 328-332.

金鹏, 王静, 朱虹, 等, 2012. 果蔬采后冷害控制技术及机制研究进展 [J]. 南京农业大学学报 (5): 171-178.

孔祥佳, 林河通, 郑俊峰, 等, 2011. 诱导冷藏橄榄果实抗冷性的适宜热空气处理条件优化 [J]. 农业工程学报, 27 (8): 371-376.

李灿婴, 侯佳宝, 张浪, 等, 2021. 外源甜菜碱处理对南果梨果实贮藏品质的影响 [J]. 包装与食品机械, 39 (4): 31-37.

李朝森, 张婷, 蒋晓红, 等, 2018. 不同贮藏方式对甜玉米可溶性糖含量的影响 [J]. 上海蔬菜 (1): 69-71.

李翠, 侯柄竹, 2023. 脱落酸调控果实成熟的分子及信号转导机制研究进展 [J]. 果树学报, 40 (5): 988-999.

李翠丹, 申琳, 生吉萍, 2013. 一氧化氮参与水杨酸诱导的采后番茄果实抗病性反应 [J]. 食品科学, 34 (8): 294-298.

李东, 郝旺, 雷雨, 等, 2024. 氯化钙/聚乙烯醇涂膜处理对脆红李生理生化的影响 [J]. 中国食品学报, 24 (2): 218-227.

李夫庆, 张子德, 李素玲, 等, 2009. 赤霉素 (GA_3) 处理对甜樱桃果实品质和采后生理的影响 [J]. 食品工业科技 (10): 301-304.

李汉良, 2011. 没食子酸丙酯对新高梨软化和褐变的影响 [J]. 农产品加工 (9): 51-53.

李欢, 张舒怡, 张钟, 等, 2017. 鲜食枣与制干枣的成熟软化机理差异研究 [J]. 西北林学院学报, 32 (5): 137-143.

李慧民, 牛建新, 党小燕, 2008. 化学药剂处理克服香梨自交不亲和性效果研究 [J]. 新疆农业科学, 45 (6): 1076-1079.

李江阔, 纪淑娟, 张鹏, 等, 2009. 南果梨褐变因子分布及贮藏过程中的变化趋势［J］. 食品研究与开发（3）：148-150.

李丽花，程曦，朱璇，2017. 振动胁迫对杏果实超微结构的影响［J］. 食品工业科技，38（16）：280-284，290.

李美玲，林育钊，王慧，等，2019. 能量状态在果蔬采后衰老中的作用及其调控研究进展［J］. 食品科学，40（9）：290-295.

李明启，1989. 果实生理［M］. 北京：科学出版社.

李囡，姜子涛，李荣，2007. 食品中没食子酸丙酯的定量分析研究进展［J］. 食品研究与开发，28（9）：172-175.

李宁，赵蕊，王越，2016. MAP 结合低温贮藏对草莓保鲜效果的影响［J］. 保鲜与加工，16（1）：12-15.

李佩艳，尹飞，苏娇，等，2022. 草酸处理减轻采后果蔬冷害机制研究进展［J］. 食品与发酵工业，48（24）：319-326.

李秋棉，罗均，李雪萍，等，2012. 果实香气物质的合成与代谢研究进展［J］. 广东农业科学，39（19）：104-107.

李双芳，刘生杰，赵胡，等，2020. 外源水杨酸处理对香椿芽采后贮藏品质的影响［J］. 食品科技，45（8）：38-43.

李斯宇，2019. 外源赤霉素处理对采后芒果果皮色泽形成的影响［D］. 海口：海南大学.

李伟明，陈晶晶，段雅婕，等，2018. 番荔枝果实后熟过程多糖代谢与果实软化和采后裂果的关系［J］. 植物生理学报，54（11）：1727-1736.

李晓晶，2023. PuWRKY74/PuNAC37 调控'南果梨'果实香气形成关键基因 PuAAT1 的分子机制［D］. 沈阳：沈阳农业大学.

李亚玲，崔宽波，石玲，等，2020. 近冰温贮藏对杏果实冷害及活性氧代谢的影响［J］. 食品科学，41（7）：177-183.

李姚瑶，2020. p-香豆酸处理调控采后桃果实冷害的机理研究［D］. 西安：西北大学.

李英华，2009. 正己醇对草莓果实采后抑菌及保鲜效果的研究［D］. 乌鲁木齐：新疆农业大学.

李圆圆，罗安伟，李琳，等，2018. 采前氯吡脲处理对'秦美'猕猴桃贮藏期间果实硬度及细胞壁降解的影响［J］. 食品科学，39（21）：273-278.

李跃，李国瑞，陈永胜，2020. 微生物代谢工程在花色苷生产过程中的应用现状和前景［J］. 食品科学，41（13）：260-266.

李正国，罗爱民，刘勤晋，2000. 钙处理对水蜜桃果实成熟的影响 [J]. 食品科学（7）：15-16.

李志欣，刘进余，岳雷，等，2019. 冬枣高效栽培措施 [J]. 中国果菜，39（11）：100-102.

李自芹，李文绮，贾文婷，等，2023. 氯化钙与1-MCP对西州蜜甜瓜采后贮藏品质的影响 [J]. 新疆农业科学，60（7）：1698-1704.

梁芳菲，王小容，邓丽莉，等，2018. 采后柑橘果实糖酸代谢研究进展 [J]. 食品与发酵工业，44（10）：268-274.

梁金甜，曾丽萍，赵文汇，等，2024. 主成分分析γ-氨基丁酸-壳聚糖涂膜对人参果贮藏品质的影响 [J]. 包装工程，45（1）：165-173.

林植芳，李双顺，林桂珠，1985. 花生离体叶片衰老的调节-Ⅰ. 抗坏血酸和甘氨酸对几种酶活性的影响 [J]. 植物生理学通讯（4）：33-35.

凌关庭，2000. 有"第七类营养素"之称的多酚类物质 [J]. 中国食品添加剂（1）：28-37.

刘彩红，2022. 正丁醇对哈密瓜果实采后抗冷性影响的研究 [D]. 乌鲁木齐：新疆农业大学.

刘恩信，马永昆，1997. 胡萝卜的贮藏与病害预防 [J]. 新疆农垦科技（3）：27.

刘剑锋，程云清，彭抒昂，2004. 梨采后细胞壁成分及果胶酶活性与果肉质地的关系 [J]. 园艺学报（5）：579-583.

刘菊华，徐碧玉，张静，等，2010. Mads-box转录因子的相互作用及对果实发育和成熟的调控 [J]. 遗传，32（9）：893-902.

刘锴栋，敬国兴，袁长春，等，2012. 外源抗坏血酸对圣女果采后生理和抗氧化活性的影响 [J]. 热带作物学报，33（10）：1851-1855.

刘琦，陈国刚，任雷厉，等，2010. 库尔勒香梨保鲜状况及贮藏中的新问题 [J]. 农产品加工·学刊（4）：73-77.

刘强，张贵友，陈受宜，2000. 植物转录因子的结构与调控作用 [J]. 科学通报，45（14）：1465-1474.

刘淑清，2011. 甜樱桃的采后处理与保鲜 [J]. 农产品加工（5）：20-21.

刘学，刘丽丹，温顺位，等，2015. 冷害对贮藏期间果蔬色泽的影响及其作用机制 [J]. 保鲜与加工，15（4）：74-76，80.

刘雪峰，向苹苹，马晓丽，等，2021. 外源糖处理对塔罗科血橙果实品质的影响 [J]. 安徽农业科学，49（9）：48-50，53.

刘雪艳，张洁仙，魏佳，等，2023.氯化钙调控能量代谢途径对小白杏采后转色的影响［J］.食品科学，44（23）：177-186.

刘拥海，俞乐，王若仲，等，2011.抗坏血酸对植物生长发育的作用及其缺失突变体的研究进展［J］.植物生理学报，47（9）：847-854.

芦玉佳，张昱，宋美玉，等，2023.外源果糖处理对采后杏果实软化的影响［J］.食品科学，44（11）：152-159.

鲁奇林，赵宏侠，冯叙桥，等，2014.气调包装贮藏对鲜枣采后贮藏生理和效果的影响［J］.食品与发酵工业，40（5）：216-221.

陆玲鸿，马媛媛，古咸彬，等，2002.猕猴桃果实软化过程中细胞壁多糖物质含量与果胶降解相关酶活性变化［J］.浙江农业学报，34（12）：2648-2658.

陆云梅，黄仁华，夏仁学，2011.红肉脐橙果实中抗氧化物质含量及其抗氧化活性变化［J］.果树学报，28（1）：134-137.

吕莹，陈芹芹，李旋，等，2023.干燥对果蔬加工色泽影响的研究进展［J］.食品科学，44（13）：368-377.

罗晓莉，高彦祥，2023.姜黄素及其微胶囊化技术研究进展［J］.中国食品添加剂，34（5）：315-330.

罗自生，2005.柿果实采后软化过程中细胞壁组分代谢和超微结构的变化［J］.植物生理与分子生物学学报（6）：651-656.

马大文，张华.茉莉酸甲酯熏蒸提高树莓果实采后贮藏品质［J］.食品研究与开发，2018，39（8）：166-169.

马德伟，1992.甜瓜优良品种［J］.中国蔬菜（5）：45.

马海珍，2018.玉米转录因子ZmbZIP4功能的研究［D］.济南：山东大学.

马丽丽，左进华，王清，等，2021.UV-C处理对青椒色泽和生理品质的影响［J］.食品科学，42（3）：281-288.

马烁，赵华，2023.果蔬汁防褐变的研究进展［J］.农产品加工（11）：76-79，83.

马玄，常雪花，郭科燕，等，2015.水杨酸处理对杏果实采后抗病性及活性氧代谢的影响［J］.食品科技（4）：57-61.

马艳艳，2018.外源水杨酸处理对甜樱桃果实品质和贮藏效果的影响［D］.郑州：河南农业大学.

玛合沙提·努尔江，包天雨，张添琪，等，2023.红曲色素的生物活性及其作用机制研究进展 [J]. 食品与发酵工业，49（6）：347-356.

孟令波，褚向明，秦智伟，等，2001. 关于甜瓜起源与分类的探讨 [J]. 北方园艺（4）：20-21.

苗红霞，金志强，刘伟鑫，等，2013. 香蕉采后果肉硬度与淀粉代谢变化 [J]. 中国农学通报，29（28）：124-128.

莫亿伟，郑吉祥，李伟才，2010. 外源抗坏血酸和谷胱甘肽对荔枝保鲜效果的影响 [J]. 农业工程学报（3）：363-367.

木合塔尔·扎热，李疆，罗淑萍，等，2012. 全光和遮光下库尔勒香梨果实品质的比较分析 [J]. 经济林研究，30（4），27-31.

潘家丽，陈舒柔，李木火，等，2023. 氯化钙处理对采后百香果细胞壁物质代谢的影响 [J]. 食品研究与开发，44（16）：32-39.

盘柳依，向妙莲，陈明，等，2019. 茉莉酸甲酯对'金魁'猕猴桃冷藏期间生理生化的影响 [J]. 分子植物育种，17（7）：2363-2370.

庞荣丽，张巧莲，郭琳琳，等，2012. 水果及其制品中果胶含量的比色法测定条件优化 [J]. 果树学报，29（2）：302-307.

彭丽桃，蒋跃明，姜微波，等，2002. 园艺作物乙烯控制研究进展 [J]. 食品科学，23（7）：132-136.

齐秀东，魏建梅，2015. 冷藏和乙烯处理对采后苹果果实糖代谢及关键基因表达的调控 [J]. 现代食品科技，31（7）：137-145.

钱丽丽，杨斯琪，李跃，等，2023. 果实采后软化机制研究进展 [J]. 食品工业科技，45（4），371.

丘智晃，冯紫荟，陈煜林，等，2022. 叶面喷施不同钙源对辣椒生长及其果实品质的影响 [J]. 福建农业学报，37（12）：1562-1570.

屈红霞，唐友林，谭兴杰，等，2001. 采后菠萝贮藏品质与果肉细胞超微结构的变化 [J]. 果树学报，18（3）：164-167.

桑月英，2022. 外源水杨酸和脱落酸调控冬枣采后抗氧化系统研究 [D]. 石河子：石河子大学.

桑兆泽，郑鄢燕，贾丽娥，等，2022. 外源水杨酸处理对青椒采后机械伤的影响 [J]. 北方园艺（20）：92-98.

邵丽梅，2023. 基于苯丙烷和活性氧代谢的蟠桃 ε-聚赖氨酸雾化保鲜调控机制研究 [D]. 沈阳：沈阳农业大学.

盛蕾，2016. 基于膜脂代谢的冷藏南果梨果皮褐变分子机制及防褐调控研究 [D]. 沈阳：沈阳农业大学.

施俊凤, 薛梦林, 王春生, 等, 2009. 甜樱桃采后生理特性与保鲜技术的研究现状与进展 [J]. 保鲜与加工, 9 (6): 7-10.

石飞, 何馨, 王君, 2023. 茶多酚海藻酸钠涂膜处理葡萄的保鲜效果 [J]. 食品研究与开发, 44 (12): 61-66.

苏青青, 2014. 富士苹果贮藏期间果实品质的研究 [D]. 杨凌: 西北农林科技大学.

孙德兰, 陈建敏, 宋艳梅, 等, 2008. 细胞膜的形貌结构及其功能信息 [J]. 中国科学, 38 (2): 101-108.

孙华军, 2020. 转录因子 PuMYB21/54 协同调控膜脂代谢关键基因 PuPLDβ1 介导冷藏南果梨果皮褐变的分子机制 [D]. 沈阳: 沈阳农业大学.

孙思胜, 王玉敏, 田林垚, 等, 2022. 采后氯化钙处理对'阳光玫瑰'葡萄贮藏品质的影响 [J]. 许昌学院学报, 41 (5): 75-79.

孙婷, 王峰, 2019. 红曲色素在食品中的应用 [J]. 农产品加工 (18): 70-72.

孙文文, 刘梦培, 纵伟, 等, 2021. 采后钙处理对甜柿贮藏品质的影响 [J]. 食品工业, 42 (11): 220-224.

孙秀兰, 刘兴华, 张华云, 等, 2001. 变温贮藏对黑琥珀李品质及生理特性的影响 [J]. 西北农林科技大学学报 (自然科学版), 29 (2): 109-113.

孙扬扬, 2020. 基于膜脂代谢的常温贮藏南果梨果心褐变机理及调控研究 [D]. 沈阳: 沈阳农业大学.

孙也, 周红利, 罗璇, 等, 2024. 气调包装对 H3 蓝莓采后贮藏品质的影响 [J]. 中国果树 (6): 57-61, 113.

谭维军, 杨军泽, 刘兴辉, 等, 2004. 大樱桃贮运保鲜技术 [J]. 西北园艺 (6): 53-54.

汤梅, 罗洁莹, 张浣悠, 等, 2018. 不同保鲜处理对鹰嘴蜜桃贮藏品质的影响 [J]. 现代食品科技, 34 (3): 167-172.

汤月昌, 许凤, 董栓泉, 等, 2015. 果糖对西兰花抗氧化性及其品质的影响 [J]. 现代食品科技, 31 (4): 164-169, 293.

汤月昌, 许凤, 王鸿飞 等, 2014. 葡萄糖处理对青花菜品质和抗氧化性的影响 [J]. 食品科学, 35 (14): 205-209.

唐继兴, 2023. 脱落酸处理调控果实采后品质机理研究 [D]. 北京: 中国农业科学院.

唐瑶, 陈洋, 曹婉鑫, 2016. 多酚类化合物的分类、来源及功能研究进展 [J]. 中国食物与营养, 22 (3): 32-34.

田龙, 2007. 黄金梨的气调贮藏保鲜试验 [J]. 农业机械学报, 38 (10): 77-79.

田梦瑶, 周宏胜, 唐婷婷, 等, 2022. 外源蔗糖处理对采后桃果皮色泽形成的影响 [J]. 食品科学, 43 (1): 177-183.

田密霞, 胡文忠, 朱蓓薇, 2008. 抗坏血酸处理对鲜切水晶梨营养成分及褐变的影响 [J]. 食品与发酵工业, 31 (1): 156-159.

田世平, 2013. 果实成熟和衰老的分子调控机制 [J]. 植物学报, 48 (5): 481-488.

拓俊绒, 闵德安, 杜新源, 2005. 富士系苹果低温气调贮藏研究 [J]. 北方果树 (5): 13-15.

王大伟, 向延菊, 2016. 采后钙处理对新疆和田地区冬枣贮藏特性的影响 [J]. 食品工业, 37 (8): 92-95.

王恒, 李梦奇, 李燊星, 等, 2023. 决明子总蒽醌提取物抗氟尿嘧啶致小鼠肝损伤的谱效关系 [J]. 南方医科大学学报, 43 (5): 825-831.

王红林, 王宇, 赵晓珍, 等, 2021. 外源NO处理对百香果采后贮藏品质的影响 [J]. 中国南方果树, 50 (5): 54-61.

王慧, 陈燕华, 林河通, 等, 2018. 纸片型1-MCP处理对安溪油柿果实采后生理和贮藏品质的影响 [J]. 食品科学, 39 (21): 253-259.

王建勋, 高疆生, 木塔里甫, 2006. 树莓生物学特性及栽培管理技术 [J]. 山西果树 (3): 25-26.

王金字, 董文宾, 杨春红, 等, 2010. 红曲色素的研究及应用新进展 [J]. 食品科技, 35 (1): 245-248.

王静, 2012. 外源抗坏血酸 (AsA) 控制采后龙眼果实果皮褐变的生理生化机制研究 [D]. 福州: 福建农林大学.

王静, 2015. 外源抗坏血酸 (AsA) 对采后猕猴桃果实生理和品质的影响 [J]. 陕西农业科学, 61 (9): 37-41.

王俊文, 2020. NO对采后树莓贮藏品质及苯丙烷和花青素代谢的影响 [D]. 泰安: 山东农业大学.

王兰菊, 胡青霞, 李靖, 2003. 甜瓜涂膜常温保鲜效果研究 [J]. 郑州轻工业学院学报 (自然科学版) (18): 68-70.

王力娜, 范术丽, 宋美珍, 等, 2010. 植物 MADS-box 基因的研究进展 [J]. 生物技术通报 (8): 12-19.

王利斌, 姜丽, 石韵, 等, 2013. 气调对豇豆贮藏期效果的影响 [J]. 食品科学, 34 (10): 313-316.

王玲利, 林春来, 韦洁敏, 等, 2018. 钙处理对黄冠梨贮藏品质的影响 [J]. 中国园艺文摘, 34 (6): 27-29.

王娜, 黎天, 尤宏争, 等, 2023. 甘肃 3 种中草药加工副产物中关键药效成分含量分析 [J]. 中国饲料 (12): 76-80.

王文举, 张亚红, 平吉成, 等, 2010. 外源抗氧化剂对高温胁迫下红地球葡萄果实日灼的影响 [J]. 北方园艺 (1): 4-6.

王学密, 李江阔, 张鹏, 等, 2008. 南果梨适宜采收成熟度的研究 [J]. 保鲜与加工 (4): 31-34.

王瑶, 罗淑芬, 胡花丽, 等, 2019. 外源 NO 处理对采后鲜莲子品质及乙烯代谢的影响 [J]. 现代食品科技, 35 (11): 100-108.

王懿, 2021. 甜菜碱处理对桃果实采后氨基酸代谢和 AsA-GSH 循环代谢的影响 [D]. 南京: 南京农业大学.

王云香, 王清, 高丽朴, 等, 2018. 外源 NO 处理对黄瓜采后生理特性的影响 [J]. 北方园艺 (18): 109-113.

魏宝东, 梁冰, 张鹏, 等, 2014. 1-MCP 处理结合冰温贮藏对磨盘柿果实软化衰老的影响 [J]. 食品科学, 35 (10): 236-240.

魏建梅, 马锋旺, 关军锋, 等, 2009. 京白梨果实后熟软化过程中细胞壁代谢及其调控 [J]. 中国农业科学, 42 (8): 2987-2996.

温广宇, 朱文学, 2003. 天然植物色素的提取与开发应用 [J]. 河南科技大学学报 (农学版) (2): 68-74.

吴敏, 陈昆松, 贾惠娟, 等, 2003. 桃果实采后软化过程中内源 IAA、ABA 和乙烯的变化 [J]. 果树学报, 20 (3): 157-160.

吴震, 别小妹, 王和福, 1997. 南果梨果实后熟过程生理生化变化的研究 [J]. 沈阳农业大学学报, 28 (2): 111-115.

席万鹏, 郁松林, 周志钦, 2013. 桃果实香气物质生物合成研究进展 [J]. 园艺学报, 40: 1679-1690.

项丽霞, 2006. 热处理、半胱氨酸、赤霉素、真空处理对莲藕品质及表皮褐变的影响 [D]. 武汉: 华中农业大学.

谢章荟, 高静, 2024. 果蔬色泽在热加工和非热加工技术中的变化研究进

展 [J/OL]. 现代食品科技, 40 (5): 299-312.

辛丹丹, 2017. 外源褪黑素处理对黄瓜采后品质影响机理及抗冷性研究 [D]. 杨凌: 西北农林科技大学.

邢军, 杨洁, 郑力, 2005. 新疆杏子分布及贮藏保鲜的可行性分析研究 [J]. 新疆大学学报: 自然科学版 (1): 79-81.

修伟业, 黎晨晨, 遇世友, 等, 2023. 类胡萝卜素生物学功能及提高其生物利用的研究进展 [J]. 食品工业科技, 44 (10): 406-415.

徐昌杰, 陈昆松, 张上隆, 等, 1997. 猕猴桃果实采后淀粉、可溶性糖含量变化及其相关酶活性的变化 [J]. 浙江农业学报, 9 (4): 215-217.

徐倩, 殷学仁, 陈昆松, 2014. 基于乙烯受体下游转录因子的果实品质调控机制研究进展 [J]. 园艺学报, 41 (9): 1913-1923.

徐思朦, 艾少杰, 薛蕾, 等, 2023. 自发气调处理对桃果实采后冷害及风味品质的调控效应 [J]. 果树学报, 40 (9): 1952-1965.

许娟, 张校立, 李鹏, 等, 2016. 没食子酸丙酯处理对库尔勒香梨果实贮藏期品质相关指标变化的影响 [J]. 新疆农业科学, 53 (1): 126-134.

杨常碧, 朱艳, 唐萍, 等, 2022. 芦荟的生物活性研究进展 [J]. 山东化工, 51 (24): 73-75.

杨方威, 段懿菲, 冯叙桥, 2016. 脱落酸的生物合成及对水果成熟的调控研究进展 [J]. 食品科学, 37 (3): 266-272.

杨菊, 程谦伟, 孟陆丽, 等, 2022. 外源抗坏血酸对香蕉保鲜和抗氧化代谢的影响 [J]. 食品研究与开发, 43 (15): 146-151.

杨娜, 王清, 郭李维, 等, 2012. 外源抗坏血酸处理对胡萝卜贮藏期间品质的影响 [J]. 中国农学通报, 28 (15): 243-248.

杨铨珍, 景绚, 1992. 树莓营养成分及果汁加工适应性研究 [J]. 中国果树 (1): 5.

杨婷, 2018. 膜脂过氧化对植物细胞的伤害 [J]. 科技与创新 (8): 61-62.

杨文慧, 黄玉咪, 徐超, 等, 2020. 氯化钙和草酸处理减轻香蕉果实贮藏冷害 [J]. 中国南方果树, 49 (5): 78-82, 86.

杨小兰, 2020. 一氧化氮对低温处理秋葵木质化进程和贮藏品质的影响 [D]. 南京: 南京农业大学.

杨艳芬, 2009. 大樱桃采后生理与贮藏保鲜技术研究进展 [J]. 北方园艺 (11): 122-124.

杨杨，申琳，生吉萍，等，2013. 外源 NO 对采后芒果果实低温胁迫后 H_2O_2 代谢的影响［J］. 中国农业科技导报（2）：131-136.

杨颖，高世庆，唐益苗，等，2009. 植物 bZIP 转录因子的研究进展［J］. 麦类作物学报，29（4）：730-737.

姚苗苗，2018. 乙烯对冷藏南果梨酯类香气合成影响的分子机理［D］. 沈阳：沈阳农业大学.

宜景宏，孙万河，张红，等，2003. 南果梨贮藏保鲜技术［J］. 北方果树（1）：4-6.

尹雪华，王凤娜，徐玉勤，等，2017. 香椿的营养保健功能及其产品的开发进展［J］. 食品工业科技，38（19）：342-345.

于继洲，郭艳，冯磊，2004. 水杨酸对果树的生理效应［J］. 烟台果树（4）：11-12.

于天颖，蔡忠杰，郭升民，等，2010. '南果梨'采后衰老机理与保鲜技术研究进展［J］. 北方果树，9（5）：1-3.

余辰，2014. γ-氨基丁酸对梨果实青霉病抗性的诱导作用及相关机理研究［D］. 杭州：浙江大学.

余东，熊丙全，曾明，等，2004. 热带保健浆果之王——西番莲［J］. 中国南方果树，33（5）：44-45.

袁洁，朱璇，逄焕明，等，2013. 外源水杨酸处理对采后杏果实抗冷性的影响［J］. 食品工业科技，34（24）：339-343.

袁莉，毕阳，李永才，等，2011. 采后赤霉素处理对低温贮藏期间枸杞鲜果腐烂的抑制和品质的影响［J］. 食品与生物技术学报，30（5）：653-656.

袁蒙蒙，高丽朴，王清，等，2012. 壳聚糖涂膜减轻黄瓜冷害的研究［J］. 湖北农业科学，51（10）：2016-2020.

袁莹，李乐，陈静霞，等，2018. 多酚类化合物的提取及功效研究进展［J］. 粮食与油脂，31（7）：15-17.

张海新，宁久丽，及华，2010. 果实采后品质和生理变化研究进展［J］. 河北农业科学，14（2）：54-56.

张海英，韩涛，许丽，等，2008. 果实的风味构成及其调控［J］. 食品科学，29（4）：464-469.

张娟，高滋艺，杨惠娟，等，2015. '秦冠'和'富士'质地差异的解剖学观察及相关酶活性研究［J］. 西北农业学报，24（10）：88-94.

张丽萍, 2013. 冷藏及 1-MCP 处理对南果梨香气代谢的影响 [D]. 沈阳: 沈阳农业大学.

张梅, 2007. 设施桃果实香气组分及相关性研究 [D]. 泰安: 山东农业大学.

张梦如, 杨玉梅, 成蕴秀, 等, 2014. 植物活性氧的产生及其作用和危害 [J]. 西北植物学报, 34 (9): 1916-1926.

张明晶, 姜微波, 徐杏连, 等, 2002. 1-甲基环丙烯对香蕉食用品质变化的影响 [J]. 食品科学, 23 (2): 126-128.

张强, 代文婷, 李冀新, 等, 2020. 糖代谢对甜瓜果实后熟软化的影响 [J]. 食品与发酵工业, 46 (1): 112-117.

张润光, 王良艳, 黄丽婉, 2011. 甜瓜贮藏保鲜技术研究进展 [J]. 保鲜与加工, 11 (1): 36-39.

张少颖, 任小林, 饶景萍, 2005. 番茄果实采后一氧化氮处理对活性氧代谢的影响 [J]. 园艺学报, 32 (5): 818-822.

张婷婷, 2020. 不同涂膜处理对新疆冬枣保鲜效果的研究 [D]. 石河子: 石河子大学.

张唯, 严成, 陈亚利, 等, 2018. 核桃青皮萘醌类色素的超高压提取工艺优化 [J]. 食品工业, 39 (8): 51-55.

张伟清, 林媚, 王天玉, 等, 2020. 柠檬精油复合涂膜对椪柑采后品质的影响 [J]. 核农学报, 34 (12): 2725-2733.

张颖, 2017. 魔芋葡甘聚糖复合涂膜对蓝莓保鲜效果的影响 [J]. 江苏农业科学, 45 (11): 146-149.

张瑜瑜, 陈泽斌, 用成健, 等, 2022. 外源水杨酸处理对蓝莓采后生理及贮藏品质的 [J]. 西南农业学报, 35 (1): 168-175.

张瑜瑜, 用成健, 刘佳妮, 等, 2022. 氯化钙处理对蓝莓果实采后生理代谢及品质的影响 [J]. 食品研究与开发, 43 (3): 87-93.

张昱, 芦玉佳, 赵亚婷, 等, 2023. 外源葡萄糖通过调控蔗糖代谢增强采后杏果实的抗冷性 [J]. 食品科学, 44 (21), 167-183.

张钊, 王野苹, 1993. 香梨品种种源问题的探讨 [J]. 果树学报 (2): 113-115.

张杼润, 张瑞杰, 赵津, 等, 2019. 24-表油菜素内酯对杏果实采后抗冷性与可溶性糖含量的影响 [J]. 食品科学, 40 (7): 198-203.

章镇, 韩振海, 2012. 果树分子生物学 [M]. 上海: 上海科学技术出版社.

赵会杰,林学梧,1992.抗坏血酸对小麦旗叶衰老进程中膜脂过氧化的影响[J].植物生理学通讯,28(5):351-352.

赵金梅,周禾,孙启忠,等,2009.植物脂肪酸不饱和性对植物抗寒性影响的研究[J].草业科学(9):133-138.

赵临强,李宁炜,武瑜,等,2021.氯化钙处理对鲜切苹果功能成分变化的影响研究[J].现代食品(11):205-210.

赵文琦,曲长福,王翠华,等,2007.树莓的营养保健价值与市场前景浅析[J].北方园艺(6):114-115.

赵颖颖,陈京京,金鹏,等,2012.低温预贮对冷藏桃果实冷害及能量水平的影响[J].食品科学,33(4):276-281.

赵云峰,林瑜,吴玲艳,2012.茄子果实采后软化过程中细胞壁组分及其降解酶活性的变化[J].食品与发酵工业,38(10):212-216.

赵云峰,郑瑞生,2010.冷害对茄子果实贮藏品质的影响[J].食品科学,31(10):321-325.

郑柯斌,司文,杨怡妍,等,2020.3种赤霉素对蓝莓果实成熟与品质的影响[J].浙江农业科学,61(3):524-527.

郑鹏蕊,李东立,付亚波,等,2022.一种自发气调包装袋对杨梅果实采后品质的影响[J].保鲜与加工,22(6):28-34.

郑秋萍,林育钊,李美玲,等,2019.果实采后软化的影响因素及抑制技术研究进展[J].亚热带农业研究,15(4):262-270.

郑永华,李三玉,席玙芳,2000.枇杷冷藏过程中果肉木质化与细胞壁物质变化的关系[J].植物生理学报(26):306-310.

周传悦,程谦伟,张定宇,等,2023.外源赤霉素处理对采后香蕉生理生化的影响[J].中国南方果树,52(2),78-83.

周春丽,吕玲琴,钟贤武,等,2012.硝普钠(SNP)对菠菜保鲜效果的影响[J].食品研究与开发,33(2):184-186.

周春丽,钟贤武,苏虎,等,2011.一氧化氮对果蔬采后保鲜机理的研究进展[J].湖北农业科学,50(10):1954-1957.

周大祥,汪开拓,匡文玲,等,2020.果糖处理对冷藏雷竹笋品质和木质化的影响及其调控机制研究[J].食品与发酵工业,46(15):175-183.

周鑫,2015.冷藏及1-MCP处理对南果梨酯类香气影响的分子机制及香气"唤醒"研究 [D]. 沈阳：沈阳农业大学.

朱佳丽，敬璞，2017. 红曲红色素稳定性研究及光热降解动力学分析 [J]. 食品与发酵科技，53（5）：49-53，79.

朱丽琴，李斌，张伟，等，2013. NO 对采后李果实保鲜效果的影响 [J]. 江西农业大学学报（6）：1157-1161.

朱绍坤，乔军，马丽，等，2024. 外源钙对巨峰葡萄裂果及其代谢物质的影响 [J]. 江苏农业科学，52（4）：163-168.

朱树华，孙丽娜，周杰，2009. 一氧化氮对猕猴桃果实营养品质和活性氧代谢的影响 [J]. 果树学报，26（3）：334-339.

朱树华，周杰，束怀瑞，等，2005. 一氧化氮延缓草莓成熟衰老的生理效应 [J]. 中国农业科学，38（7）：1418-1424.

朱雪静，2018. 红树莓气调保鲜实验及传热传质研究 [D]. 哈尔滨：哈尔滨商业大学.

庄晓红，刘声远，马岩松，等，2008. 常温条件下南果梨主要营养成分及其变化规律的研究 [J]. 保鲜与加工（2）：34-37.

左玉，2013. 多酚类化合物研究进展 [J]. 粮食与油脂，26（4）：6-10.

ABEDI-FIROOZJAH R, PARANDI E, HEYDARI M, et al., 2023. Betalains as promising natural colorants in smart/active food packaging [J]. Food Chemistry, 424（30）：136408.

ABOUSALHAM A, NARI J, TEISSÈRE M, et al., 1997. Study of fatty acid specificity of sunflower phospholipase D using detergent phospholipids micelles [J]. Biochemistry, 248, 374-379.

ADHIKARY T, GILL P S, JAWANDHA S K, et al., 2020. Browning and quality management of pear fruit by salicylic acid treatment during low temperature storage [J]. Journal of the Science of Food & Agriculture, 101：853-862.

AGHDAM M S, FARD J R, 2017. Melatonin treatment attenuates postharvest decay and maintains nutritional quality of strawberry fruits Fragaria × anannasa cv. Selva by enhancing GABA shunt activity [J]. Food Chemistry, 221：1650-1657.

AGHDAM M S, HASSANPOURAGHDAM M B, PALIYATH G, et al., 2012. The language of calcium in postharvest life of fruits, vegetables and flowers [J]. Scientia Horticulturae, 144：102-115.

AGHDAM M S, JANNATIZADEH A, NOJADEH M S, et al., 2019b. Exogenous melatonin ameliorates chilling injury in cut anthurium flowers during low temperature storage [J]. Postharvest Biology and Technology, 148: 184-191.

AGHDAM M S, LUO Z, JANNATIZADEH A, et al., 2019a. Employing exogenous melatonin applying confers chilling tolerance in tomato fruits by up-regulating ZAT2/6/12 giving rise to promoting endogenous polyamines, proline, and nitric oxide accumulation by triggering arginine pathway activity [J]. Food Chemistry, 275: 549-556.

AGHDAM M S, NADERI R, SARCHESHMEH M A A, et al., 2015. Amelioration of postharvest chilling injury in anthurium cut flowers by γ-aminobutyric acid (GABA) treatments [J]. Postharvest Biology and Technology, 110: 70-76.

AIMAN Z, SAJAD A W, TASHOOQ A B, et al., 2022. Preparation of a biodegradable chitosan packaging film based on zinc oxide, calcium chloride, nano clay and poly ethylene glycol incorporated with thyme oil for shelf-life prolongation of sweet cherry [J]. International Journal of Biological Macromolecules, 217: 572-582.

AKTARUZZAMAN M, AFROZ T, KIM B S. et al., 2017. Occurrence of postharvest gray mold rot of sweet cherry due to Botrytis cinerea in Korea [J]. Journal of Plant Diseases & Protection, 124 (1): 93-96.

ALI S, KHAN A S, MALIK A U, et al., 2018. Pre-storage methionine treatment inhibits postharvest enzymatic browning of cold stored 'Gola' litchi fruit [J]. Postharvest Biology and Technology, 140: 100-106.

ALI Z M, CHIN L H, LAZAN H, 2004. A comparative study on wall degrading enzymes, pectin modifications and softening during ripening of selected tropical fruits [J]. Plant Science, 167 (2): 317-327.

ALTISENT R, ECHEVERRI G, GRAELL J, et al., 2009. Lipoxygenase activity is involved in the regeneration of volatile ester-synthesizing capacity after ultra-low oxygen storage of 'Fuji' apple [J]. Journal of Agricultural and Food Chemistry, 57 (10): 4305-4312.

ALVAREZBUYLLA E R, PELAZ S, LILJEGREN S J, et al., 2000. An ancestral MADS-box gene duplication occurred before the divergence of plants and animals [J]. Proceedings of the National Academy of Sciences of the United States of America, 97 (10): 5328-5333.

AMAKI K, SAITO E, TANIQUCHI K, et al., 2011. Role of chlorogenic acid quinine and interaction of chlorogenic acid quinone and catechins in the enzymatic browning of apple [J]. Bioscience Biotechnology and Biochemistry, 75: 829-832.

AMORIM I S, ALMEIDA M C S, CHAVES R P F, et al., 2022. Technological applications and color stability of carotenoids extracted from selected Amazonian fruits [J]. Food Science and Technology, 42: e01922.

ARSHAD M O, CHAUHAN Y, SINGH P, et al., 2022. Advancements in controlled atmosphere storage technology-A review [J]. Proceedings of Second International Conference in Mechanical and Energy Technology, 290: 399-410.

ASGARIAN, Z S, KARIMI R, GHABOOLI M, et al., 2022. Biochemical changes and quality characterization of cold-stored 'Sahebi' grape in response to postharvest application of GABA [J]. Food Chemistry, 373: 131401.

ASGHARI M, AGHDAM M S, 2010. Impact of salicylic acid on postharvest physiology of horticultural crops [J]. Trends in Food Science and Technology, 21: 502-509.

ASLAM M M, KOU M, DOU Y, et al., 2024. The Transcription Factor MiMYB8 Suppresses Peel Coloration in Postharvest 'Guifei' Mango in Response to High Concentration of Exogenous Ethylene by Negatively Modulating MiPAL1 [J]. International Journal of Molecular Sciences, 25 (9): 4841.

ATCHLEY W R, FITCH W M, 1997. A natural classification of the basic helix-loop-helix class of transcription factors [J]. Proceedings of the National Academy of Sciences of the United States of America, 94 (10): 5172-5176.

ATKINSON R G, GUNASEELAN K, WANG M Y, et al., 2011. Dissecting the role of climacteric ethylene in kiwifruit (Actinidia chinensis) ripening using a 1-aminocyclopropane-1-carboxylic acid oxidase knockdown line [J]. Journal of Experimental Botany, 62 (11): 3821-3835.

AYUB R, GUIS M, BEN A M, et al., 1996. Expression of ACC oxidase antisense gene inhibits ripening of cantaloupe melon fruits [J]. Nature Biotechnology, 14 (7): 862-866.

BAJWA V S, SHUKLA M R, SHERIF S M, et al., 2014. Role of melatonin in alleviating cold stress in Arabidopsis thaliana [J]. Journal of Pineal Research, 56: 238-245.

BARMAN K, ASREY R, PAL R K, 2011. Putrescine and carnauba wax pretreatments alleviate chilling injury, enhance shelf life and preserve pomegranate fruit quality during cold storage [J]. Scientia Horticulturae, 130: 795-800.

BEAUDRY R, 1999. Effect of O_2 and CO_2 partial pressure on selected phenomena affecting fruit and vegetable quality [J]. Postharvest Biology and Technology, 15 (3): 293-303.

BELAY Z A, CALEB O J, MAHAJAN P V, et al., 2018. Pomegranate arils ('Wonderful') tolerance to low O_2 stress during active modified atmosphere storage: based on real time respiration rate [J]. Acta Horticulturae Sinica, 1201: 213-220.

BELIGNI M V, LAMATTINA L, 2000. Nitric oxide stimulates seed germination and de-etiolation, and inhibits hypocotyl elongation, three light-inducible responses in plants [J]. Planta, 210 (2): 215-221.

BEN A R, KISLER N, FRENKEL C, 1993. Degradation and solubilization of pectin by β-galactosidases purified from avocado mesocarp [J]. Plant Physiol, 87: 279-285.

BERNARDO P, MARIA C, 2021. Innovative Preservation Technology for the Fresh Fruit and Vegetables [J]. Foods, 10, 719.

BHUSHAN B, KUMAR S, MAHAWAR M K, et al., 2019. Nullifying phosphatidic acid effect and controlling phospholipase D associated browning in litchi pericarp through combinatorial application of hexanal and inositol [J]. Scientific Reports, 9 (1): 2402.

BOURNE M C, 2006. Selection and use of postharvest technologies as a component of the food chain [J]. Journal of Food Science, 69 (2): 43-46.

BRAGA A R C, MURADOR D C, DE SOUZA M L M, et al., 2018. Bioavailability of anthocyanins: gaps in knowledge, challenges and future research [J]. Journal of Food Composition and Analysis, 68: 31-40.

BRAINARD G C, HANIFIN J P, GREESON J M, et al., 2001. Action spectrum for melatonin regulation in humans: Evidence for a novel circadian photoreceptor [J]. Journal of Neuroscience, 21: 6405-6412.

BRIZZOLARA S, MANGANARIS G A, FOTOPOULOS V, et al., 2020. Primary metabolism in fresh fruits during storage [J]. Frontiers in Plant Science, 11: 80.

BRUGGER P, MARKTL W, HEROLD M, 1995. Impaired nocturnal secretion of melatonin in coronary heart disease [J]. The Lancet (8962): 345.

BRUMMELL D A, 2006. Cell wall disassembly in ripening fruit [J]. Functional Plant Biology, 33 (2): 103-119.

BRUMMELL D A, 2006. Cell wall disassembly in ripening fruit [J]. Funct Plant Biol, 33 (2): 103-119.

BRUMMELL D A, BOWEN J K, GAPPER N E, 2022. Biotechnological approaches for controlling postharvest fruit softening [J]. Current Opinion in Biotechnology, 78: 102786.

BRUMMELL D A, CIN V D, LURIE S, et al., 2004. Cell wall metabolism during the development of chilling injury in cold-stored peach fruit: Association of mealiness with arrested disassembly of cell wall pectins [J]. Journal of Experimental Botany, 55 (405): 2041-2052.

BU J, YU Y, AISIKAER G, et al., 2013. Postharvest UV-C irradiation inhibits the production of ethylene and the activity of cell wall-degrading enzymes during softening of tomato (*Lycopersicon esculentum* L.) fruit [J]. Postharvest Biology and Technology (86): 337-345.

BURKHARDT S, TAN D X, MANCHESTER L C, et al., 2001. Detection and quantification of the antioxidant melatonin in Montmorency and balaton tart cherries (Prunus cerasus) [J]. Journal of Agricultural and Food Chemistry, 49: 4898-4902.

BYEON Y, BACK K, 2016. Low melatonin production by suppression of either serotonin N-acetyltransferase or N-acetylserotonin methyltransferase in rice causes seedling growth retardation with yield penalty, abiotic stress susceptibility, and enhanced coleoptile growth under anoxic conditions [J]. Journal of Pineal Research, 60 (3): 348-359.

CAI C, XU C J, SHAN L L, et al., 2006. Low temperature conditioning reduces postharvest chilling injury in loquat fruit [J]. Postharvest Biology and Technology, 41: 252-259.

CAI J, MO X, WEN C, et al., 2021. FvMYB79 positively regulates strawberry fruit softening via transcriptional activation of FvPME38 [J]. International journal of molecular sciences, 23 (1): 101.

CALVO-BRENES P, O'HARE T, 2020. Effect of freezing and cool storage on carotenoid content and quality of zeaxanthin-biofortified and standard yellow sweet-corn (*Zea mays* L.) [J]. Journal of Food Composition and Analysis, 86: 103353.

CAO H, CHEN J, YUE M, et al., 2020. Tomato transcriptional repressor MYB70 directly regulates ethylene-dependent fruit ripening [J]. Plant Journal, 104: 1568-1581.

CAO S F, SONG C B, SHAO J R, et al., 2016. Exogenous melatonin treatment increases chilling tolerance and induces defense response in harvested peach fruit during cold storage [J]. Journal of Agricultural and Food Chemistry, 64 (25): 5215-5222.

CARDINALI D P, BRUSCO L I, LIBERCZUK C, et al., 2002. The use of melatonin in Alzheimer's disease [J]. Neuroendocrinol. Lett, 23: 20-23.

CARRASCO-ORELLANA C, STAPPUNG Y, MENDEZ-YAÑEZ A, et al., 2018. Characterization of a ripeningrelated transcription factor FcNAC1 from Fragaria chiloensis fruit [J]. Scientific Reports, 8 (1): 1-12.

CECCARELLI A, FARNETI B, KHOMENKO I, et al., 2020. Nectarine volatilome response to fresh-cutting and storage [J]. Postharvest Biology and Technology, 159, 111020.

CEZAR D R N A, MARASCHIN M, DI PIERO R M, 2015. Antifungal activity of salicylic acid against penicillium expansum and its possible mechanisms of action [J]. International Journal of Food Microbiology, 215: 64-70.

CHARAN S M, GOMEZ K B, SHEEIA P B, et al., 2017. Effect of storage conditions and duration on quality of passion fruit (*Passiflora edulis* sims.) nectar [J]. Asian Journal of Dairy and Food Research, 36 (2): 161-165.

CHEN C J, SUN C C, WANG Y H, et al., 2023. The preharvest and postharvest application of salicylic acid and its derivatives on storage of fruit and vegetables: A review [J]. Scientia Horticulturae, 312, 111858.

CHEN D, SHAO Q, YIN L, et al., 2019. Polyamine function in plants: metabolism, regulation on development, and roles in abiotic stress responses [J]. Frontiers in Plant Science, 9, 1945. https://doi.org/10.3389/fpls.2018.01945.

CHEN H, CAO S, JIN Y, et al., 2016. The Relationship between CmADHs and the diversity of volatile organic compounds of three aroma types of melon (Cucumis melo) [J]. Frontiers in Physiology, 7: 254.

CHEN H, LIANG J, WU F, et al., 2013. Effect of hypobaric storage on flesh lignification, active oxygen metabolism and related enzyme activities in bamboo shoots [J]. LWT-Food Science and Technology, 51 (1): 190-195.

CHEN J, MAO L, LU W, et al., 2016. Transcriptome profiling of postharvest strawberry fruit in response to exogenous auxin and abscisic acid [J]. Planta, 243 (1): 183-197.

CHEN L L, SHAN W, CAI D L, et al., 2021. Postharvest application of glycine betaine ameliorates chilling injury in cold-stored banana fruit by enhancing antioxidant system [J]. Scientia Horticulturae, 287: 110264.

CHEN Y, HUNG Y C, CHEN M, et al., 2017. Effects of acidic electrolyzed oxidizing water on retarding cell wall degradation and delaying softening of blueberries during postharvest storage [J]. LWT-Food Science and Technology, 84: 650-657.

CHEN Y, SUN J, LIN H, et al., 2017. Paper-based 1-MCP treatment suppresses cell wall metabolism and delays softening of Huanghua pears during storage [J]. Journal of the Science of Food and Agriculture, 97 (8): 2547-2552.

CHENG G P, YANG E, LU W J, 2009. Effect of nitric oxide on ethylene synthesis and softening of banana fruit slice during ripening [J]. Journal of Agricultural and Food Chemsitry, 57 (13): 5799-5804.

CHERVIN C, SPEIRS J, LOVEYS B, et al., 2000. Influence of low oxygen storage on aroma compounds of whole pears and crushed pear flesh [J]. Postharvest Biology and Technology, 19 (3): 279-285.

CHOI H R, BAEK M W, CHEOL L H, et al., 2022. Changes in metabolites and antioxidant activities of green 'Hayward' and gold 'Haegeum' kiwifruits during ripening with ethylene treatment [J]. Food Chemistry, 384, 132490.

CHOI H R, BAEK M W, TILAHUN S, et al., 2022. Long-term cold storage affects metabolites, antioxidant activities, and ripening and stress-related genes of kiwifruit cultivars [J]. Postharvest Biology and Technology, 189, 111912.

CHONG J X, LAI S, YANG H, 2015. Chitosan combined with calcium chloride impacts fresh-cut honeydew melon by stabilising nanostructures of sodium-carbonate-soluble pectin [J]. Food Control, 53: 195-205.

CHOUDHURY S R, PANDEY S, 2017. Phosphatidic acid binding inhibits rgs1 activity to affect specific signaling pathways in Arabidopsis [J]. The Plant Journal, 90 (3): 466-477.

CHRISTIAN C, SPEIRS J, LOVEYS B R, et al., 2000. Influence of low oxygen storage on aroma compounds of whole pears and crushed pear flesh [J]. Postharvest Biology and Technology, 19: 279-285.

CHUNG M Y, VREBALOV J, ALBA R, et al., 2010. A tomato (Solanum lycopersicum) APETALA2/ERF gene, SlAP2a, is a negative regulator of fruit ripening [J]. The Plant Journal, 64 (6): 936-947.

CIERESZKO I, 2018. Regulatory roles of sugars in plant growth and development [J]. Acta Societatis Botanicorum Poloniae, 87 (2): 3583.

CONN E E. 1984. Chemical conjugation and compartmentation: plant adaptations to toxic natural products [J]. Cellular and Molecular Biology of Plant Stress, 351-365.

CONTADOR L, SHINYA P, INFANTE R, 2015. Texture phenotyping in fresh fleshy fruit [J]. Scientia Horticulturae, 193: 40-46.

CONTRERAS C, TJELLSTRÖM H, BEAUDRY R M, 2015. Relationships between free and esterified fatty acids and lox-derived volatiles during ripening in apple [J]. Postharvest Biology and Technology, 112: 105-113.

COSGROVE D J, 2014. Re-constructing our models of cellulose and primary cell wall assembly [J]. Current Opinion In Plant Biology, 22: 122-131.

COUEE I, SULMON C, GOUESBET G, et al., 2006. Involvement of soluble sugars in reactive oxygen species balance and responses to oxidative stress in plants [J]. Journal of Experimental Botany, 57 (3): 449-459.

DE BRUXELLES G L, ROBERTS M R, 2001. Signals Regulating Multiple Responses to Wounding and Herbivores [J]. Critical Reviews in Plant Sciences, 20, 487-521.

DEBEAUJON I, PEETERS A J M, LE'OM-KLOOSTERZIEL K M, et al., 2001. The *TRANSPARENT TESTA12* gene of Arabidopsis encodes a multidrug secondary transporter-like protein required for flavonoid sequestration in vacuoles of the seed coat endothelium [J]. Plant Cell, 13: 853-872.

DEBRA M D, LAURENCE D M, CHRISTOPHER B W, 1992. Cell wall changes in nectarines (Prunus persica): Solubilization and depolymerization of pectic and neutral polymers during ripening and in mealy fruit [J]. Plant Physiology, 100 (3): 1203-1210.

DEFILIPPI B G, MANRÍQUEZ D, LUENGWILAI K, et al., 2009. Chapter 1 aroma volatiles: biosynthesis and mechanisms of modulation during fruit ripening [J]. Advances in Botanical Research, 2009, 50: 1-37.

DEK, MOHD, SABRI, et al., 2018. Inhibition of tomato fruit ripening by 1-MCP, wortmannin and hexanal is associated with a decrease in transcript levels of phospholipase D and other ripening related genes [J]. Postharvest Biology and Technology, 140: 50-59.

DENG J, ZHOU Y, BAI M, et al., 2010. Anxiolytic and sedative activities of *Passiflora edulis* f. flavicarpa [J]. Journal of Ethnopharmacology, 128 (1), 148-153.

DENG Y, XU L, ZENG X, et al., 2010. New perspective of GABA as an inhibitor of formation of advanced lipoxidation end-products: it's interaction with malondiadehyde [J]. Journal of biomedical nanotechnology 6: 318-324.

DHALL R K, 2013. Advances in Edible Coatings for Fresh Fruits and Vegetables: A Review [J]. Critical Reviews in Food Science and Nutrition, 53: 435-450.

DHAWAN K, SHARMA A, 2002. Antitussive activity of the methanol extract of passiflora incarnata leaves [J]. Fitoterapia, 73 (5): 397-399.

DHONUKSHE P, LAXALT A M, GOEDHART J, et al., 2003.

Phospholipase D activation correlates with microtubule reorganization in living plant cells. Plant Cell Online, 15, 2666-2679.

DI H, ZHANG Y, MA J, et al., 2022. Sucrose treatment delays senescence and maintains the postharvest quality of baby mustard (Brassica junceavar. gemmifera) [J]. Food Chemistry, 14: 100272.

DOLLINS A B, ZHDANOVA I V, WURTMAN R J, et al., 1994. Effect of inducing nocturnal serum melatonin concentrations in daytime on sleep, mood, body temperature, and performance [J]. Proceedings of the National Academy of Sciences of the United States of America, 91: 1824-1828.

DUBBELS R, REITER R J, KLENKE E, et al., 1995. Melatonin in edible plants identified by radioimmunoassay and by high-performance liquid chromatography-mass spectrometry [J]. Journal of Pineal Research, 18: 28-31.

DUNLOP M J, DOSSANI Z Y, SZMIDT H L, et al., 2014. Engineering microbial biofuel tolerance and export using efflux pumps [J]. Molecular System Biology, 7 (1): 487.

DUVAL M, HSIEH T F, KIM S Y, et al., 2002. Molecular characterization of AtNAM: a member of the Arabidopsis NAC domain superfamily [J]. Plant Molecular Biology, 50 (2): 237-248.

ELBAGOURY M M, TUROOP L, RUNO S, et al., 2020. Regulatory influences of methyl jasmonate and calcium chloride on chilling injury of banana fruit during cold storage and ripening [J]. Food Science & Nutrition, 9 (2): 929-942.

ELITZUR T, YAKIR E, QUANSAH L, et al., 2016. Banana MaMADS transcription factors are necessary for fruit ripening and molecular tools to promote Shelf-Life and food security [J]. Plant Physiology, 171: 380-391.

FAN X G, ZHAO H D, WANG X M, et al., 2017. Sugar and organic acid composition of apricot and their contribution to sensory quality and consumer satisfaction [J]. Scientia Horticulturae, 225: 553-560.

FAN X, JIANG W, GONG H, et al., 2019. Cell wall polysaccharides degradation and ultrastructure modification of apricot during storage at a near freezing temperature [J]. Food Chemistry, 300: 125194.

FAN X, SHU C, ZHAO K, et al., 2018. Regulation of apricot ripening and softening process during shelf life by post-storage treatments of exogenous ethylene and 1-methylcyclopropene [J]. Scientia Horticulturae, 232, 63-70.

FAN Z Q, BA L J, SHAN W, et al., 2018. A banana R2R3-MYB transcription factor MaMYB3 is involved in fruit ripening through modulation of starch degradation by repressing starch degradation-related genes and MabHLH6 [J]. Plant Journal, 96 (6): 1191-1205.

FELLMAN, J. K, 2000. Factors that influence biosynthesis of volatile flavor compounds in apple fruits [J]. Horticultural Science, 35 (6): 1026-1033.

FICKER E, TAGLIALATELA M, WIBLE B A, et al., 1994. Spermine and spermidine as gating molecules for inward rectifier K^+ channels [J]. Science, 266: 1068-1072.

FISCHER R L, BENNETT A B, 1991. Role of cell wall hydrolases in fruit ripening [J]. Annual Review of Plant Biology, 42: 675-703.

FRANCISCOB F B, PALOMA S B, MONIKA V, et al., 2008. Effects of a pretreatment with nitric oxide on peach (*Prunus persica* L.) storage at room temperature [J]. European Food Research and Technology, 227 (6): 1599-1611.

FRIED R, OPREA I, FLECK K, et al., 2022. Biogenic colourants in the textile industry-a promising and sustainable alternative to synthetic dyes [J]. Green Chemistry, 24 (1): 13-35.

FU C C, HAN C, WEI Y X, et al., 2024. Two NAC transcription factors regulated fruit softening through activating xyloglucan endotransglucosylase/hydrolase genes during kiwifruit ripening [J]. International Journal of Biological Macromolecules, 263, 130678.

FU C, CHEN H, GAO H, et al., 2020. Two papaya MYB proteins function in fruit ripening by regulating some genes involved in cell-wall degradation and carotenoid biosynthesis [J]. Journal of the Science of Food and Agriculture, 100 (12): 4442-4448.

FZA B, AJA B, FAB K, et al., 2019. Effect of methyl jasmonate on wound healing and resistance in fresh-cut potato cubes [J]. Postharvest Biology and Technology, 157: 110958.

GALSTON A W, KAUR-SAWHNEY R, DAVIS P J, 1995. Polyamines as endogenous growth regulators, Plant Hormones: Physiology, Biochemistry and Molecular Biology [J]. Kluwer Academic Publishers, 158-178.

GAN Z, YUAN X, SHAN N, et al., 2021. AcERF1B and AcERF073 Positively Regulate Indole-3-acetic Acid Degradation by Activating AcGH 3.1 Transcription during Postharvest Kiwifruit Ripening [J]. Journal of Agricultural and Food Chemistry, 69 (46): 13859-13870.

GAN Z, YUAN X, SHAN N, et al., 2021. AcWRKY40 mediates ethylene biosynthesis during postharvest ripening in kiwifruit [J]. Plant Science, 309: 110948.

GAO H, LU Z M, YANG Y, et al., 2018. Melatonin treatment reduces chilling injury in peach fruit through its regulation of membrane fatty acid contents and phenolic metabolism [J]. Food Chemistry, 245: 659-666.

GAO Y, KAN C N, WAN C P, et al., 2018. Effects of hot air treatment and chitosan coating on citric acid metabolism in ponkan fruit during cold storage [J]. Plos One, 13 (11): 0206585.

GARCIA-CAYUELA T, GOMEZ DE CADINANOS L P, PELAEZ C, et al., 2012. Expression in Lactococcus lactis of functional genes related to amino acid catabolism and cheese aroma formation is influenced by branched chain amino acids [J]. International Journal of Food Microbiology. 159: 207-213.

GE H, ZHANG J, ZHANG Y J, et al., 2017. EjNAC3 transcriptionally regulates chilling-induced lignification of loquat fruit via physical interaction with an atypical CAD-like gene [J]. Journal of Experimental Botany, 68 (8): 5129-5136.

GHOSH A, SAHA I, DEBNATH S C, et al., 2021. Chitosan and putrescine modulate reactive oxygen species metabolism and physiological responses during chili fruit ripening [J]. Plant Physiology and Biochemistry, 163: 55-67.

GIOVANNONI J, NGUYEN C, AMPOFO B, et al., 2017. The epigenome and transcriptional dynamics of fruit ripening [J]. Annual Review of Plant Biology, 68: 61-84.

GOLDENBERG L, YANIV Y, CHOI H J, et al., 2016. Elucidating the biochemical factors governing off-flavor perception in mandarins [J]. Postharvest Biology and Technology, 120, 167-179.

GONDA I, BAR E, PORTNOY V, et al., 2010. Branched-chain and aromatic amino acid catabolism into aroma volatiles in *Cucumis melo* L. fruit [J]. Journal of Experimental Botany, 61 (4): 1111-1123.

GONZALEZ E A, AGRASAR A T, CASTRO L M P, et al., 2011. Solid-state fermentation of red raspberry (*Rubus ideaus* L.) and arbutus berry (*Arbutus unedo* L.) and characterization of their distillates [J]. Food Research International, 44 (5), 1419-1426.

GOODMAN C D, CASATI P, WALBOT V, 2004. A multidrug resistance-associated protein involved in anthocyanin transport in Zea mays [J]. Plant Cell, 16: 1812-1826.

GOULAO L F, OLIVEIRA C M, 2008. Cell wall modifications during fruit ripening: When a fruit is not the fruit [J]. Trends in Food Science & Technology, 19: 4-25.

GU R, ZHU S, ZHOU J, et al., 2014. Inhibition on brown rot disease and induction of defence response in harvested peach fruit by nitric oxide solution [J]. European Journal of Plant Pathology, 139 (2): 369-378.

GU S, XU D, ZHOU F, et al., 2022. Repairing ability and mechanism of methyl jasmonate and salicylic acid on mechanically damages sweet cherries [J]. Sciential Horticulturae, 292: 110567.

GUO A Y, XIN C, GE G, et al., 2008. PlantTFDB: a comprehensive plant transcription factor database [J]. Nucleic Acids Research, 36: 966.

GUO S L, SONG J, ZHANG B B, et al., 2018. Genome-wide identification and expression analysis of betagalactosidase family members during fruit softening of peach [J]. Postharvest Biology And Technology, 136: 111-123.

GUO X, LI Q, LUO T, et al., 2023. Postharvest calcium chloride treatment strengthens cell wall structure to maintain litchi fruit quality [J]. Foods, 12 (13): 2478.

HABIBI F, VALERO D, SERRANO M, et al., 2022. Exogenous Application of Glycine Betaine Maintains Bioactive Compounds, Antioxidant Activity, and Physicochemical Attributes of Blood Orange Fruit During Prolonged Cold

Storage [J]. Frontiers in Nutrition, 9: 873915.

HADI M E, AHMED M, ZHANG F J, et al., 2013. Advances in fruit aroma volatile research [J]. Molecules, 18 (7): 8200-8229.

HAN A, CAO S F, LI Y X, et al., 2019. Sucrose treatment suppresses programmed cell death in broccoli florets by improving mitochondrial physiological properties [J]. Postharvest Biology and Technology, 156: 110932.

HAN S K, NAN Y Y, QU W, et al., 2018. Exogenous γ-aminobutyric acid treatment that contributes to regulation of malate metabolism and ethylene synthesis in apple fruit during storage [J]. Journal of Agriculture and Food Chemistry, 66 (51): 13473-13482.

HAN S, LIU H, HAN Y, et al., 2021. Effects of calcium treatment on malate metabolism and γ-aminobutyric acid (GABA) pathway in postharvest apple fruit [J]. Food chemistry, 334: 127479.

HAN Y, EAST A, NICHOLSON S, et al., 2022. Benefits of modified atmosphere packaging in maintaining 'Hayward' kiwifruit quality at room temperature retail conditions [J]. New Zealand Journal of Crop and Horticultural Science, 50 (2-3): 242-258.

Harb, J., BISHARAT R, STREIF J, 2008. Changes in volatile constituents of blackcurrants (*Ribes nigrum* L. cv. 'Titania') following controlled atmosphere storage [J]. Postharvest Biology and Technology, 47 (3): 271-279.

HARDELAND R, 2016. Melatonin in plants-Diversity of levels and multiplicity of functions [J]. Frontiers in Plant Science, 7: doi: 10.3389/fpls.2016.00198.

HASSAN I, ZHANG Y, DU G, et al., 2007. Effect of salicylic acid (SA) on delaying fruit senescence of Huang Kum pear [J]. Frontiers of Agriculture in China, 1: 456-459.

HATTORI A, MIGITAKA H, IIGO M, et al., 1995. Identification of melatonin in plants and its effects on plasma melatonin levels and binding to melatonin receptors in vertebrates [J]. Biochemistry and Molecular Biology International, 35: 627-634.

HAYAT S, ALI B, AHMAD A, 2007. Salicylic acid: Biosynthesis, metabolism and physiological role in plants [M]. Dordrecht: Springer.

HERNÁNDEZ I G, GOMEZ F J V, CERUTTI S, et al., 2015. Melatonin in Arabidopsis thaliana acts as plant growth regulator at low concentrations and preserves seed viability at high concentrations [J]. Plant Physiology and Biochemistry, 94: 191-196.

HINCHA D K, HOFNER R, SCHWAB K B, et al., 1987. Membrane rupture is the common cause of damage to chloroplast membranes in leaves injured by freezing or excessive wilting [J]. Plant Physiology, 83: 251-253.

HOLLAND N, MENEZES H C, LAFUENTE M T, 2002. Carbohydrates as related to the heat-induced chilling tolerance and respiratory rate of 'Fortune' mandarin fruit harvested at different maturity stages [J]. Postharvest Biology and Technology, 25: 181-191.

HOLLAND N, NUNES F L D S, DE MEDEIROS I U D, et al., 2012. Hightemperature conditioning induces chilling tolerance in mandarin fruit: A cell wall approach [J]. Journal of The Science of Food and Agriculture, 92, 3039-3045.

HONG H T, PHAN A D T, O'HARE T J, 2021. Temperature and maturity stages affect anthocyanin development and phenolic and sugar content of purple–pericarp supersweet sweetcorn during storage [J]. Journal of Agricultural and Food Chemistry, 69 (3): 922-931.

HOU H, KONG X, ZHOU Y, et al., 2022. Genome-wide identification and characterization of bZIP transcription factors in relation to litchi (Litchi chinensis Sonn.) fruit ripening and postharvest storage [J]. International Journal of Biological Macromolecules, 222: 2176-2189.

HOU Q, UFER G, BARTELS D, 2016. Lipid signaling in plant responses to abiotic stress [J]. Plant Cell and Environment, 39: 1029-1048.

HU L Y, HU S L, WU J, et al., 2012. Hydrogen sulfide prolongs postharvest shelf life of strawberry and plays an antioxidative role in fruits [J]. Journal of Agricultural and Food Chemistry, 60 (35): 8684-8693.

HU M J, ZHU Y Y, LIU G S, et al., 2019. Inhibition on anthracnose and induction of defense response by nitric oxide in pitaya fruit [J]. Scientia horticulturae, 245: 224-230.

HU Z L, DENG L, CHEN X Q, et al., 2010. Co-suppression of the EIN2-

homology gene LeEIN2 inhibits fruit ripening and reduces ethylene sensitivity in tomato [J]. Russian Journal of Plant Physiology, 57: 554-559.

HUANG R, LI C Y, GUO M, et al., 2022. Caffeic acid enhances storage ability of apple fruit by regulating fatty acid metabolism [J]. Postharvest Biology and Technology, 92: 112012.

HUANG S, BI Y, LI H, et al., 2023. Reduction of Membrane Lipid Metabolism in Postharvest Hami Melon Fruits by n-Butanol to Mitigate Chilling Injury and the Cloning of Phospholipase D-β Gene [J]. Foods, 12: 1904.

HUANG W, SHI Y, YAN H, et al., 2023. The calcium-mediated homogalacturonan pectin complexation in cell walls contributes the firmness increase in loquat fruit during postharvest storage [J]. Journal of Advanced Research, 49: 47-62.

HUANG Y W, NIE Y X, WAN Y Y, et al., 2013. Exogenous glucose regulates activities of antioxidant enzyme, soluble acid invertase and neutral invertase and alleviates dehydration stress of cucumber seedlings [J]. Scientia Horticulturae, 162: 20-30.

HUANG Z, LI J, ZHANG J, et al., 2017. Physicochemical properties enhancement of Chinese kiwi fruit (Actinidia chinensis Planch) via chitosan coating enriched with salicylic acid treatment [J]. Journal of Food Measurement and Characterization, 11 (1): 184-191.

IGBAVBOA U, HAMILTON J, KIM H Y, et al., 2002. Anew role for apolipoprotein E: Modulating transport of polyunsaturated phospholipid molecular species in synaptic plasma membranes [J]. Journal of Neurochemistry, 80, 255-261.

IRELAND H, YAO J, TOMES S, et al., 2013. Apple SEPALLATA1/2-like genes control fruit flesh development and ripening [J]. Plant Journal, 73, 1044-1056.

ITO Y, KITAGAWA M, IHASHI N, et al., 2008. DNA-binding specificity, transcriptional activation potential, and the rin mutation effect for the tomato fruit-ripening regulator RIN [J]. Plant Journal, 55 (2): 212-223.

IZAWA T, FOSTER R, CHUA N H, 1993. Plant bZIP protein DNA binding specificity [J]. Journal of Molecular Biology, 230: 1131-1144.

JAIN V, CHAWLA S, CHOUDHARY P, et al., 2019. Post-harvest calcium chloride treatments influence fruit firmness, cell wall components and cell wall hydrolyzing enzymes of Ber (Ziziphus mauritiana Lamk.) fruits during storage [J]. Journal of Food Science and Technology, 56 (10): 4535-4542.

JALALIA P, ZAKERINA A R, ABOUTALEBI-JAHROMIA A H, et al., 2023. Improving postharvest life, quality and bioactive compounds of strawberry fruits using spermine and spermidine [J]. Brazilian Journal of Biology, 83: e273886.

JANNATIZADEH A, AGHDAM M S, LUO Z S, et al., 2019. Impact of Exogenous Melatonin Application on Chilling Injury in Tomato Fruits During Cold Storage [J]. Food and Bioprocess Technology, 12: 741-750.

JARISARAPURIN W, SANRATTANA W, CHULAROJMONTRI L, et al., 2019. Antioxidant Properties of Unripe Carica papaya Fruit Extract and Its Protective Effects against Endothelial Oxidative Stress [J]. Evidence-Based Complementary and Alternative Medicine, 1-15.

JI Y, HU W, LIAO J, et al., 2021. Ethanol vapor delays softening of postharvest blueberry by retarding cell wall degradation during cold storage and shelf life [J]. Postharvest Biology and Technology, 177: 111538.

JIA H F, WANG S S, LIN H, et al., 2018. Effects of abscisic acid agonist or antagonist applications on aroma volatiles and anthocyanin biosynthesis in grape berries [J]. The Journal of Horticultural Science and Biotechnology, 93 (4): 392-399.

JIA H, XIE Z, WANG C, et al., 2017. Abscisic acid, sucrose, and auxin coordinately regulate berry ripening process of the Fujiminori grape [J]. Functional and Integrative Genomics, 17: 441-457.

JIA L, CLEGG M T, JIANG T, 2004. Evolutionary dynamics of the DNA binding domains in putative R2R3-MYB genes identified from rice subspecies indica and japonica genomes [J]. Plant Physiology, 134: 575-585.

JIA Y, LI W, 2015. Characterisation of lipid changes in ethylene-promoted senescence and its retardation by suppression of phospholipase Dδ in Arabidopsis leaves [J]. Frontiers in Plant Science, 6: 1045.

JIANG P K, XU Q F, XU Z H, et al., 2006. Seasonal changes in soil labile organic carbon pools within a Phyllostachys praecox stand under high rate fertilization and winter mulch in subtropical China [J]. Forest Ecology and Management, 236 (1): 30-36.

JIANG Y M, JIANG Y L, QU H X, et al., 2007. Energy aspects in ripening and senescence of harvested horticultural crops [J]. Stewart Postharvest Review, 3 (2): 1-5.

JIANG Y M, SONG L L, LIU H, et al., 2006. Postharvest characteristics and handling of litchi fruit-an overview [J]. Australian Journal of Experimental Agriculture, 46: 1541-1556.

JIN L F, CAI Y T, SUN C, et al., 2019. Exogenous L-glutamate treatment could induce resistance against Penicillium expansum in pear fruit by activating defense-related proteins and amino acids metabolism [J]. Postharvest Biology and Technology, 150: 148-157.

JIN P, ZHU H, WANG L, et al., 2014. Oxalic acid alleviates chilling injury in peach fruit by regulating energy metabolism and fatty acid contents [J]. Food Chemistry, 161: 87-93.

JIN Y R X, LI C Y, ZHANG S R, et al., 2024. Sucrose, cell wall, and polyamine metabolisms involve in preserving postharvest quality of 'Zaosu' pear fruit by L-glutamate treatment [J]. Plant Physiology and Biochemistry, 208: 108455.

JOHNSTON S J, CARROLL J S, 2015. Transcription factors and chromatin proteinsas therapeutic targets in cancer [J]. Biochimica Et Biophysica Acta, 1855 (2): 183-192.

JONES J A, VERNACCHIO V R, SINKOE A L, et al., 2016. Experimental and computational optimization of an Escherichia coli co-culture for the efficient production of flavonoids [J]. Metabolic Engineering, 35: 55-63.

JOYCE C P, SAMUEL C, CECIL, S, 2005. Relationship of cold acclimation, total phenolic content and antioxidant capacity with chilling tolerance in petunia (Petunia× hybrida) [J]. Environmental and Experimental Botany, 53 (2): 225-232.

KAKKAR R K, SAWHNEY V K, 2002. Polyamine research in plants-a changing perspective [J]. Physiologia Plantarum, 116: 281-292.

KALDE M, BARTH M, SOMSSICH I E, et al., 2003. Members of the Arabidopsis WRKY group III transcription factors are part of different plant defense signaling pathways [J]. Molecular Plant-microbe Interactions: MPMI, 16 (4): 295-305.

KAMDEE C, IMSABAI W, KIRK R, et al., 2014. Regulation of lignin biosynthesis in fruit pericarp hardening of mangosteen (*Garcinia mangostana* L.) after impact [J]. Postharvest Biology and Technology, 97: 68-76.

KANT K, ARORA A, SINGH V P, 2016. Salicylic acid influences biochemical characteristics of harvested tomato (*Solanum lycopersicon* L.) during ripening [J]. Indian Journal of Plant Physiology, 21 (1): 50-55.

KATAGIRI T, TAKAHASHI S, SHINOZAKI K, 2010. Involvement of a novel arabidopsis phospholipase D, AtPLDδ, in dehydration - inducible accumulation of phosphatidic acid in stress signalling [J]. The Plant Journal, 26 (6): 595-605.

KAUR-SAWHNEY R, APPLEWHITE P B, 1993. Endogenous protein bound polyamines: correlation with regions of cell division in tobacco leaves, internodes and ovaries [J]. Plant Growth Regulation, 12: 223-227.

KAWAIDE H, 2006. RIVIEW: Biochemical and molecular analyses of gibberellin biosynthesis in fungi [J]. Bioscience Biotechnology and Biochemistry, 70 (3): 583-590.

KEVANY B M, TIEMAN D M, TAYLOR M G, et al., 2007. Ethylene receptor degradation controls the timing of ripening in tomato fruit [J]. Plant Journal, 51: 458-467.

KHALIL U, RAJWANA I A, RAZZAQ K, et al., 2024. Evaluation of modified atmosphere packaging system developed through breathable technology to extend postharvest life of fresh muscadine berries [J]. Food Science and Nutrition, 12: 3663-3673.

KHAN N, FATIMA F, HAIDER M S, et al., 2019. Genome-Wide identification and expression profiling of the polygalacturonase (PG) and pectin methylesterase (PME) genes in grapevine (*Vitis vinifera* L.) [J]. International Journal of Molecular Sciences, 20 (13): 3180.

KHOSROSHAHI M R Z, ESNA-ASHARI, 2017. Postharvest putrescine treatments extend the storage life of apricot (*Prunus armeniaca* L.) Tokhm-sefid fruit [J]. Journal of Horticultural Science and Biotechnology, 82 (6):

986-990.

KIM J-S, LEE J, EZURA H, 2022. SlMBP3 Knockout/down in Tomato: Normal-Sized Fruit with Increased Dry Matter Content through Non-Liquefied Locular Tissue by Altered Cell Wall Formation [J]. Plant and Cell Physiology, 63 (10): 1485-1499.

KINNERSLEY A M, TURANO F J, 2000. Gamma aminobutyric acid (GABA) and plant responses to stress [J]. Critical Reviews in Plant Sciences, 19: 479-509.

KISHITANI S, WATANABE K, YASUDA S, et al., 1994. Accumulation of glycinebetaine during cold acclimation and freezing tolerance in leaves of winter and spring barley plants [J]. Plant Cell and Environment, 17 (1): 89-95.

KLEMPNAUER K H, GONDA T J, MICHAEL B J, 1982. Nucleotide sequence of the retroviral leukemia gene v-myb and its cellular progenitor c-myb: the architecture of a transduced oncogene [J]. Cell, 31 (2): 453-463.

KOBYLI'NSKA A, BOREK S, POSMYK M M, 2018. Melatonin redirects carbohydrates metabolism during sugar starvation in plant cells [J]. Journal of Pineal Research, 64.

KONG X M, WEI B D, GAO Z, et al., 2018. Changes in membrane lipid composition and function accompanying chilling injury in bell peppers [J]. Plant Cell and Physiology, 59 (1): 167-178.

KOU X, CHAI L, YANG S, et al., 2020. Physiological and metabolic analysis of winter jujube after postharvest treatment with calcium chloride and a composite film [J]. Journal of the Science of Food and Agriculture, 101 (2): 703-717.

KOUSHESH S M, MORADI S, 2017. Sodium nitroprusside (snp) spray to maintain fruit quality and alleviate postharvest chilling in jury of peach fruit [J]. Scientia Horticulturae, 216: 193-199.

KRAUSE G H, GRAFFLAGE S, RUMICH-BAYER S, et al., 1988. Effects of freezing on plant mesophyll cells [J]. Symposia of the Society for Experimental Biology, 42: 311-327.

KRUPA T, LATOCHA P, LIWI'NSKA A, 2011. Changes of physicochemical quality, phenolics and vitamin C content in hardy kiwifruit (Actinidia arguta and its hybrid) during storage [J]. Scientia Horticulturae, 130 (2): 410-417.

KUMAR R, TAMBOLI V, SHARMA R, et al., 2018. NAC-NOR mutations in tomato Penjar accessions attenuate multiple metabolic processes and prolong the fruit shelf life [J]. Food Chem, 259: 234-244.

KUNERT K J, EDERER M, 1985. Leaf aging, lipid peroxidation: The role of the antioxidant vitamin C and E [J]. Physical plant (65): 85-88.

LAMPUGNANI E R, KHAN G A, SOMSSICH M, et al., 2018. Building a plant cell wall at a glance [J]. Journal of Cell Science, 131 (2): s207373.

LARA I, MIRÓ R M, FUENTES T, et al., 2003. Biosynthesis of volatile aroma compounds in pear fruit stored under long-term controlled-atmosphere conditions [J]. Postharvest Biology and Technology, 29: 29-39.

LASTDRAGER J, HANSON J, SMEEKENS S, 2014. Sugar signals and thecontrol of plant growth and development [J]. Journal of Experimental Botany, 65 (3): 799-807.

LATOCHA P, KRUPA T, WOŁOSIAK R, et al., 2010. Antioxidant activity and chemical difference in fruit of different Actinidia sp [J]. International Journal of Food Sciences and Nutrition, 61 (4): 381-394.

LEI X Y, ZHU R Y, ZHANG G Y, et al., 2004. Attenuation of cold-induced apoptosis by exogenous melatonin in carrot suspension cells: The possible involvement of polyamines [J]. Journal of Pineal Research, 36: 126-131.

LEIN W, SAALBACH G, 2001. Cloning and direct G-protein regulation of phospholipase D from tobacco [J]. Biochimica et Biophysica Acta, 1530: 172-183.

LEJA M, MARECZEK A, BEN J, 2003. Antioxidant properties of two apple cultivars during long-term storage [J]. Food Chemistry, 80: 303-307.

LEONARD E, YAN Y, FOWLER Z L, et al., 2008. Strain improvement of recombinant Escherichia coli for efficient production of plant flavonoids [J]. Molecular Pharmaceutics, 5: 257-265.

LEONARD E, YAN Y, KOFFAS M A G, 2006. Functional expression of a

P450 flavonoid hydroxylase for the biosynthesis of plant-specific hydroxylated flavonols in Escherichia coli [J]. Metabolic Engineering, 8: 172-181.

LEONARD W, ZHANG P, YING D, et al., 2022. Surmounting the off-flavor challenge in plant-based foods [J]. Critical Reviews in Food Science and Nutrition, 1-22.

LERNER A B, CASE J D, TAKAHASHI Y, et al., 1958. Isolation of melatonin, a pineal factor that lightens melanocytes [J]. Journal of American Chemical Society, 80: 2587-2587.

LESHEM Y A Y, PINCHASOV Y, 2000. Non-invasive photoacoustic spectroscopic determination of relative endogenous nitric oxide and ethylene content stoichiometry during the ripening of strawberries Fragaria anannasa (Duch.) and avocados Persea americana (Mill.) [J]. Journal of Experimental Botany, 51: 1471-1473.

LESHEM Y Y, HARAMATY E, 1996. The characterization and contrasting effects of the nitric oxide free radical in vegetative stress and senescence of pisum sativum linn. foliage [J]. Journal of Plant Physiology, 148 (3-4): 258-263.

LI A, CHEN J, LIN Q, et al., 2021. Transcription Factor MdWRKY32 Participates in Starch-Sugar Metabolism by Binding to the MdBam5 Promoter in Apples During Postharvest Storage [J]. Journal of Agricultural and Food Chemistry, 69 (49): 14906-14914.

LI C Y, ZHANG C Y, LIU J X, et al., 2023. L-Glutamate maintains the quality of apple fruit by mediating carotenoid, sorbitol and sucrose metabolisms [J]. Journal of the Science of Food and Agriculture, 103 (10): 4944-4955.

LI C Y, ZHANG J H, GE Y H, et al., 2020. Postharvest acibenzolar-S-methyl treatment maintains storage quality and retards softening of apple fruit [J]. Journal of Food Biochemistry, 44: e13141.

LI C, WANG P, WEI Z, et al., 2012. The mitigation effects of exogenous melatonin on salinity-induced stress in Malus hupehensis [J]. Journal of Pineal Research, 53: 298-306.

LI D, LI L, XIAO G N, et al., 2018. Effects of elevated CO_2 on energy metabolism and γ-aminobutyric acid shunt pathway in postharvest strawberry fruit [J]. Food Chemistry, 265: 281-289.

LI D, LIMWACHIRANON J, LI L, et al., 2016. Involvement of energy metabolism to chilling tolerance induced by hydrogen sulfide in cold-stored banana fruit [J]. Food Chemistry, 208: 272-278.

LI J J, LI F, QIAN M, et al., 2017. Characteristics and regulatory pathway of the Prupe SEP1 SEPALLATA gene during ripening and softening in peach fruits [J]. Plant Science, 257: 63-73.

LI J X, ZHOU X, WEI B D, et al., 2019. GABA application improves the mitochondrial antioxidant system and reduces peel browning in 'Nanguo' pears after removal from cold storage [J]. Food Chemistry, 297: 124903.

LI J, LUO M, ZHOU X, et al., 2021. Polyamine treatment ameliorates pericarp browning in cold-stored 'Nanguo' pears by protecting mitochondrial structure and function [J]. Postharvest Biology and Technology, 178 (8): 111553.

LI J, TAO X, LI L, et al., 2016. Comprehensive RNA-Seq analysis on the regulation of tomato ripening by exogenous auxin [J]. PLOS One, 11 (5): e0156453.

LI L, KITAZAWA H, WANG X Y, et al., 2017. Regulation of respiratory pathway and electron transport chain in relation to senescence of postharvest white mushroom (Agaricus bisporus) under high O_2/CO_2 controlled atmospheres [J]. Journal of Agricultural and Food Chemistry, 65 (16): 3351-3359.

LI L, LI J, SUN J, et al., 2019. Role of phospholipase D inhibitor in regulating expression of senescence related phospholipase D gene in postharvest longan fruit [J]. Current Bioinformatics, 14: 649-657.

LI L, LV F Y, GUO Y Y, et al., 2016. Respiratory pathway metabolism and energy metabolism associated with senescence in postharvest broccoli (*Brassica oleracea* L. Var. Italica) florets in response to O_2/CO_2 controlled atmospheres [J]. Postharvest Biology and Technology, 111: 330-336.

LI P, ZHENG X, LIU Y, et al., 2014. Pre-storage application of oxalic acid alleviates chilling injury in mango fruit by modulating proline metabolism and energy status under chilling stress [J]. Food Chemistry, 142: 72-78.

LI S, CHEN K, GRIERSON D, 2019. A critical evaluation of the role of ethylene and MADS transcription factors in the network controlling fleshy fruit ripening [J]. New Phytologist, 221 (4): 1724-1741.

LI T, JIANG Z Y, ZHANG, L C, et al., 2016. Apple (Malus domestica) MdERF2 negatively affects ethylene biosynthesis during fruit ripening by suppressing MdACS1 transcription [J]. Plant Journal, 88 (5): 735-748.

LI X X, DUAN X P, JIANG H X, et al., 2006. Genome-wide analysis of basic/helixloop-helix transcription factor family in rice and Arabidopsis [J]. Plant Physiology, 141 (4): 1167-1184.

LI X, XU C J, KORBAN S S, et al., 2010. Regulatory mechanisms of textural changes in ripening fruits [J]. Critical Reviews in Plant Sciences, 29: 222-243.

LI Y, ZHU B, XU W, et al., 2007. LeERF1 positively modulated ethylene triple response on etiolated seedling, plant development and fruit ripening and softening in tomato [J]. Physiology and Biochemistry, 26 (11): 1999-2008.

LI Z, YU J, PENG Y, et al., 2016. Metabolic pathways regulated by γ-aminobutyric acid (GABA) contributing to heat tolerance in creeping bentgrass (Agrostis stolonifera) [J]. Scientific Reports, 6: 30338.

LI Y, QI H, JIN Y, et al., 2016. Role of ethylene in biosynthetic pathway of related-aroma volatiles derived from amino acids in oriental sweet melons (Cucumis melo var. makuwa Makino) [J]. Scientia Horticulturae, 201: 24-35.

LIAO G, LIU Q, XU X, et al., 2021. Metabolome and Transcriptome Reveal Novel Formation Mechanism of Early Mature Trait in Kiwifruit (Actinidia eriantha) [J]. Frontiers in Plant Science, 12, 760496.

LICAUSI F, OHME T M, PERATA P, 2013. APETALA2/Ethylene Responsive Factor (AP2/ERF) transcription factors: mediators of stress responses and developmental programs [J]. New Phytologist, 199 (3): 639-649.

LIM C G, FOWLER Z L, HUELLER T, et al., 2011. High-yield resveratrol production in engineered Escherichia coli [J]. Applied and Environment Microbiology, 77: 3451-3460.

LIM C G, WONG L, BHAN N, et al., 2015. Development of a recombinant escherichia coli strain for overproduction of the plant pigment anthocyanin [J]. Applied and Environmental Microbiology, 81 (18): 6276-6284.

LIN Y F, HU Y H, LIN H T, et al., 2013. Inhibitory effects of propyl gallate on tyrosinase and its application in controlling pericarp browning of harvested longan fruits [J]. Journal of Agricultural & Food Chemistry, 61 (11): 2889-2895.

LIN Y F, LIN H T, LIN Y X, et al., 2016. The roles of metabolism of membrane lipids and phenolics in hydrogen peroxide-induced pericarp browning of harvested longan fruit [J]. Postharvest Biology and Technology, 111: 53-61.

LIN Y F, LIN Y X, LIN H T, et al., 2017. Hydrogen peroxide-induced pericarp browning of harvested longan fruit in association with energy metabolism [J]. Food Chemistry, 225: 31-36.

LIN Y, LIN L, LIN Y, et al., 2023. Comparison between two cultivars of longan fruit cv. 'Dongbi' and 'Fuyan' in the metabolisms of lipid and energy and its relation to pulp breakdown [J]. Food Chemistry, 398: 133885.

LIN Y, LIN Y, LIN H, et al., 2018. Effects of paper containing 1-MCP postharvest treatment on the disassembly of cell wall polysaccharides and softening in Younai plum fruit during storage [J]. Food Chemistry, 264: 1-8.

LIN Y, LIN Y, LIN Y, et al., 2019. A novel chitosan alleviates pulp breakdown of harvested longan fruit by suppressing disassembly of cell wall polysaccharides [J]. Carbohydrate Polymers, 217: 126-134.

LIN Z, ZHONG S, GRIERSON D, 2009. Recent advances in ethylene research [J]. Journal of Experimental Botany, 60 (12): 3311-3336.

LIPSICK J S, 1996. One billion years of MYB [J]. Oncogene, 13 (2): 223-235.

LIU F, DOU T X, HU C H, et al., 2023. WRKY transcription factor MaWRKY49 positively regulates pectate lyase genes during fruit ripening of *Musa acuminata* [J]. Plant Physiology and Biochemistry, 194: 643-650.

LIU G S, LI H L, GRIERSON D, et al., 2022. NAC transcription factor family regulation of fruit ripening and quality: a review [J]. Cell, 11

(3): 525.

LIU H, LI D L, XU W C, et al., 2021. Application of passive modified atmosphere packaging in the preservation of sweet corns at ambient temperature [J]. LWT-Food Science and Technology, 136: 110295.

LIU H, SONG L, YOU Y, et al., 2011. Cold storage duration affects litchi fruit quality, membrane permeability, enzyme activities and energy charge during shelf time at ambient temperature [J]. Postharvest Biology and Technology, 60 (1): 24-30.

LIU J, LI Q, CHEN J, et al., 2020. Revealing Further Insights on Chilling Injury of Postharvest Bananas by Untargeted Lipidomics [J]. Foods, 9: 894.

LIU L, WHITE M J, MACRAE T H, 1999. Transcription factors and their genes in higher plants functional domains, evolution and regulation [J]. European Journal of Biochemistry, 262 (2): 247-257.

LIU M, CHEN Y, CHEN Y, et al., 2018. The tomato ethylene response factor Sl-ERF. B3 integrates ethylene and auxin signaling via direct regulation of Sl-Aux/IAA27 [J]. New Phytologist, 219 (2): 631-640.

LIU M, ZHANG Z, XU Z, et al., 2021. Overexpression of SlMYB75 enhances resistance to Botrytis cinerea and prolongs fruit storage life in tomato [J]. Plant Cell Reports, 40: 43-58.

LIU T, WANG H, KUANG J, et al., 2015. Short-term anaerobic, pure oxygen and refrigerated storage conditions affect the energy status and selective gene expression in litchi fruit [J]. LWT-Food Science and Technology, 60 (2): 1254-1261.

LIU R L, WANG Y Y, QIN G Z, et al., 2016. Molecular basis of 1-methylcyclopropene regulating organic acid metabolism in apple fruit during storage [J]. Postharvest Biology and Technology, 117: 57-63.

LU X, SUN D, LI Y, et al., 2011. Pre-and post-harvest salicylic acid treatments alleviate internal browning and maintain quality of winter pineapple fruit [J]. Scientia Horticulturae, 130 (1): 97-101.

LU Z M, WANG X L, CAO M M, et al., 2019. Effect of 24-epibrassinolide on sugar metabolism and delaying postharvest senescence of kiwifruit during ambient storage [J]. Scientia Horticulturae, 253: 1-7.

LUENGWILAI K, BECKLES D M, PLUEMJIT O, et al., 2014. Postharvest quality and storage life of 'Makapuno' coconut (*Cocos nucifera* L.) [J]. Scientia Horticulturae, 175: 105-110.

LUO M L, GE W Y, SUN H J, et al., 2022. Salicylic acid treatment alleviates diminished ester production in cold-stored 'Nanguo' pear by promoting the transcription of PuAAT [J]. Postharvest Biology and Technology, 187: 111849.

LUO M, ZHOU X, SUN H, et al., 2020. Glycine Betaine treatment alleviates loss of aroma-related esters in cold-stored 'Nanguo' pears by regulating the lipoxygenase pathway [J]. Food Chemistry, 316: 126335.

LURIE S, CRISOSTO C H, 2005. Chilling injury in peach and nectarine [J]. Postharvest Biology and Technology, 37 (3): 195-208.

LURIE S, KLEIN J D, ARIE R B, 1991. Prestorage heat treatment delays development of superficial scald on Granny Smith apple [J]. Hort Science, 26 (2): 166-167.

LYNCH K M, ZANNINI E, WILKINSON S, et al., 2019. Physiology of acetic acid bacteria and their role in vinegar and fermented beverages [J]. Comprehensive Reviews in Food Science and Food Safety, 18 (3): 587-625.

LYONS J M, 1973. Chilling injury in plants [J]. Annual Review of Plant Physiology, 24: 445-466.

MADANI B, MIRSHEKARI A, YAHIA E, 2015. Effect of calcium chloride treatments on calcium content, anthracnose severity and antioxidant activity in papaya fruit during ambient storage [J]. Journal of the Science of Food and Agriculture, 96 (9): 2963-2968.

MADEBO M P, SI-MING L, LI W, et al., 2021. Melatonin treatment induces chilling tolerance by regulating the contents of polyamine, γ-aminobutyric acid, and proline in cucumber fruit [J]. Journal of Integrative Agriculture, 20 (11): 3060-3074.

MALEKZADEH P, KHOSRAVI-NEJAD F, HATAMNIA A A, et al., 2017. Impact of postharvest exogenous γ-aminobutyric acid treatment on cucumber fruit in response to chilling tolerance [J]. Physiology and Molecular Biology of Plants, 23: 827-836.

MANCHESTER L C, TAN D X, REITER R J, et al., 2000. High levels of melatonin in the seeds of edible plants-Possible function in germ tissue protection [J]. Life Sciences, 67: 3023-3029.

MANGANARIS G A, VASILAKAKIS M, DIAMANTIDIS G, et al., 2007. The effect of postharvest calcium application on tissue calcium concentration, quality attributes, incidence of flesh browning and cell wall physicochemical aspects of peach fruits [J]. Food Chemistry, 100 (4): 1385-1392.

MARE C, MAZZUCOTELLI E, CROSATTI C, et al., 2004. Hv-WRKY38: a new transcription factor involved in cold-and drought-response in barley [J]. Plant Molecular Biology, 55 (3): 399-416.

MARQUES D N, REIS S P, DESOUZA C R B, 2017. Plant NAC transcription factors responsive to abiotic stresses [J]. Plant Gene, 11: 170179.

MARRS K A, ALFENITO M R, LLOYD A M, et al., 1995. A glutathione S-transferase involved in vacuolar transfer encoded by the maize gene Bronze-2 [J]. Nature, 1995, 375: 397-400.

MARTINEZ V, NIEVES-CORDONES M, LOPEZ-DELACALLE M, et al., 2018. Tolerance to stress combination in tomato plants: New Insights in the protective role of melatonin [J]. Molecules, 23: 535.

MARTÍN-PIZARRO C, VALLARINO J G, OSORIO S, et al., 2021. The NAC transcription factor FaRIF controls fruit ripening in strawberry [J]. The Plant Cell, 33 (5): 1574-1593.

MATHEW A G, PARPIA H A B, 1971. Food browning as a polyphenol reaction [J]. Advance in Food Research, 19: 75-145.

MATTUS A E, STAPPUNG Y, HERRERA R, et al., 2023. Molecular actors involved in the softening of fragaria chiloensis fruit accelerated by ABA treatment [J]. Journal of Plant Growth Regulation, 42: 433-448.

MAYER, A. M. 1987. Polyphenoloxidase in plants-recent progress [J]. Phytochemistry (26): 11-20.

MIGICOVSKY Z, YEATS T H, WATTS S, et al., 2021. Apple ripening is controlled by a NAC transcription factor [J]. Frontiers in Genetics, 12: 908.

MIRANDA S, VILCHES, P, SUAZO M, et al., 2020. Melatonin triggers metabolic and gene expression changes leading to improved quality traits of two sweet cherry cultivars during cold storage [J]. Food Chemistry, 319: 126360.

MIRDEHGHAN S H, RAHIMI S, 2016. Pre-harvest application of polyamines enhances antioxidants and table grape (*Vitis vinifera* L.) quality during postharvest period [J]. Food Chemisty, 196: 1040-1047.

MIRSHEKARI A, MADANI B, YAHIA E M, et al., 2019. Postharvest melatonin treatment reduces chilling injury in sapota fruit [J]. Journal of the Science of Food and Agriculture, 100 (5): 10198.

MISHRA S, BARMAN K, SINGH A K, et al., 2022. Exogenous polyamine treatment preserves postharvest quality, antioxidant compounds and reduces lipid peroxidation in black plum fruit [J]. South African Journal of Botany (146): 662-668.

MIZRAHI Y, DOSTAL H C, CHERRY J H, 1976. Protein differences between fruits of rin, a non-ripening tomato mutant, and a normal variety [J]. Planta, 130 (2): 223-224.

MO Y W, GONG D, LIANG G B, et al., 2008. Enhanced preservation effects of sugar apple fruits by salicylic acid treatment post-harvest storage [J]. Journal of the Science of Food and Agriculture, 88: 2693-2699.

MOLAEI S, RABIEI V, SOLEIMANI A, et al., 2021. Exogenous application of glycine betaine increases the chilling tolerance of pomegranate fruits cv. malase Saveh during cold storage [J]. Journal of Food Processing and Preservation, 45: e15315.

MORALES M D, GONZÁLEZ M C, REVIEJO A J, et al., 2005. A composite amperometric tyrosinase biosensor for the determination of the additive propyl gallate in foodstuffs [J]. Microchemical Journal, 80 (1): 71-78.

MORIOKA S, UEKI J, KOMARI T, 1997. Characterization of two distinctive genomic clones (accession nos. AB001919 and AB001920) for phospholipase D from rice (PGR 97-076) [J]. Plant Physiology, 114: 396.

MORRIS A J, ENGEBRECHT J, FROHMAN M A, 1996. Structure and regulation of phospholipase D [J]. Trends in Pharmacology Science, 17:

182-185.

MUNNIK T, IRVINE R F, MUSGRAVE A, 1998. Phospholipid signaling in plants [J]. Biochimica et Biophysica Acta, 1389: 222-272.

MUZAMMIL H, MUHAMMAD I H, MUHAMMAD U G, 2015. Salicylic acid induced resistance in fruits to combat against postharvest pathogens: a review [J]. Archives of Phytopathology & Plant Protection, 48 (1): 34-42.

NAKANO T, SUZUKI K, FUJIMURA T, et al., 2006. Genome-wide analysis of the ERF gene family in Arabidopsis and rice [J]. Plant Physiology, 140 (2): 411-432.

NAKATA Y, IZUMI H, 2020. Microbiological and quality responses of strawberry fruit to high CO_2 controlled atmosphere and modified atmosphere storage [J]. HortScience, 55 (3): 386-391.

NARDOZZA S, BOLDINGH H L, OSORIO S, et al., 2013. Metabolic analysis of kiwifruit (Actinidia deliciosa) berries from extreme genotypes reveals hallmarks for fruit starch metabolism [J]. Journal of Experimental Botany, 64 (16): 5049-5063.

NELSON K M, DAHLIN J L, BISS J, et al., 2017. The essential medicinal chemistry of curcumin: miniperspective [J]. Journal of Medicinal Chemistry, 60 (5): 1620-1637.

NGUYEN T B T, KETSA S, VAN DOORN W G, 2003. Relationship between browning and the activities of polyphenoloxidase and phenylalanine ammonia lyase in banana peel during low temperature storage [J]. Postharvest Biology and Technology, 30: 187-193.

NIAZI Z, RAZAVI F, KHADEMI O, et al., 2021. Exogenous application of hydrogen sulfide and γ-aminobutyric acid alleviates chilling injury and preserves quality of persimmon fruit (Diospyros kaki, cv. Karaj) during cold storage [J]. Scientia Horticulturae, 285: 110198.

NUEZ-GÓMEZ V, BAENAS N, NAVARRO-GONZÁLEZ I, et al., 2020. Seasonal variation of health-promoting bioactives in broccoli and methyljasmonate pre-harvest treatments to enhance their contents [J]. Foods, 9 (10): 1371.

NUNES M C N, EMOND J P, BRECHT J K, et al., 2007. Quality curves for mango fruit (cv. Tommy Atkins and Palmer) stored at chilling and nonchilling temperatures [J]. Journal of Food Quality, 30: 104-120.

OBENLAND D, ARPAIA M L, 2019. Effect of harvest date on off-flavor development in mandarins following postharvest wax application [J]. Postharvest Biology and Technology, 149: 1-8.

OGATA K, KANEI-ISHII C, SASAKI M, et al., 1996. The cavity in the hydrophobic core of MYB DNA-binding domain is reserved for DNA recognition and trans-activation [J]. Nature Structure and Molecular Biology, 3 (2): 178-187.

OLSEN A N, ERNST H A, LEGGIO L L, et al., 2005. NAC transcription factors: structurally distinct, functionally diverse [J]. Trends in Plant Science, 10 (2): 79-87.

OZTURK B, UZUN S, KARAKAYA O, 2019. Combined effects of aminoethoxyvinylglycine and MAP on the fruit quality of kiwifruit during cold storage and shelf life [J]. Scientia Horticulturae, 251: 209-214.

PABO C O, SAUER R T, 1992. Transcription factors: structural families and principles of DNA recognition [J]. Annual Review of Biochemistry, 61: 1053-1095.

PALMA F, CARVAJAL F, JAMILENA M, et al., 2014. Contribution of polyamines and other related metabolites to the maintenance of zucchini fruit quality during cold storage [J]. Plant Physiology and Biochemistry, 82: 161-171.

PALMA F, CARVAJAL F, JIMÉNEZ-MUÑOZ R, et al., 2019. Exogenous γ-aminobutyric acid treatment improves the cold tolerance of zucchini fruit during postharvest storage [J]. Plant Physiology et Biochemistry, 136: 188.

PALTA J P, WHITAKER B D, WEISS L S, 1993. Plasma membrane lipids associated with genetic variability in freezing tolerance and cold acclimation of Solanum species [J]. Plant Physiology, 103: 793-803.

PAN Y, ZHANG S, YUAN M, et al., 2019. Effect of glycine betaine on chilling injury in relation to energy metabolism in papaya fruit during cold storage [J]. Food Science & Nutrition, 7 (3): 1123-1130.

PANNITTERI C, CONTINELLA A, LO C L, et al., 2017. Influence of postharvest treatments on qualitative and chemical parameters of tarocco blood orange fruits to be used for fresh chilled juice [J]. Food Chemistry, 230: 441-447.

PAREDES S D, KORKMAZ A, MANCHESTER L C, et al., 2009. Phytomelatonin: a review [J]. Journal of Experimental Botany, 60: 57-69.

PAREEK S, VALERO D, SERRANO M, 2015. Postharvest biology and technology of pomegranate [J]. Journal of the Science of Food and Agriculture, 95 (12): 360-2379.

PARK E J, JEKNIC Z, CHEN T H, 2006. Exogenous application of glycinebetaine increases chilling tolerance in tomato plants (lycopersicon esculentum) [J]. Plant and Cell Physiology, 47 (6): 706-714.

PARK Y B, COSGROVE D J, 2012. A revised architecture of primary cell walls based on biomechanical changes induced by substrate-specific endoglucanases [J]. Plant Physiology, 158: 1933-1943.

PARK Y B, COSGROVE D J, 2015. Xyloglucan and its interactions with other components of the growing cell wall [J]. Plant Cell Physiology, 56: 180-194.

PATEL N, GANTAIT S, PANIGRAHI J, 2018. Extension of postharvest shelf-life in Capsicum annuum L. using exogenous application of polyamines (spermidine and putrescine), Food Chemistry, 9: 154.

PAYASI A, MISHRA N N, CHAVES A L, et al., 2009. Biochemistry of fruit softening: An overview [J]. Physiology & Molecular Biology of Plants, 15: 103-113.

PAZARES J, GHOSAL D, WIENAND U, et al., 1987. The regulatory c1 locus of zea mays encodes a protein with homology to MYB proto-oncogene products and with structural similarities to transcriptional activators [J]. Embo Journal, 6 (12): 3553-3558.

PENG Z Z, LIU G S, LI H L, et al., 2022. Molecular and genetic events determining the softening of fleshy fruits: a comprehensive review [J]. International Journal of Molecular Sciences, 23: 12482.

PESARESI P, MIZZOTTI C, COLOMBO M, et al., 2014. Genetic regulation and structural changes during tomato fruit development and ripening [J]. Frontiers in Plant Science, 5: 124.

PESIS E, 2005. The role of the anaerobic metabolites, acetaldehyde and ethanol, in fruit ripening, enhancement of fruit quality and fruit deterioration [J]. Postharvest Biology and Technology, 37 (1): 1-19.

PIDDOCK L J V, 2006. Clinically relevant chromosomally encoded multidrug resistance efflux pumps in bacteria [J]. Clinical Microbiology Reviews, 19: 382-402.

PIERI C, MARRA M, MORONI F, et al., 1994. Melatonin: A peroxyl radical scavenger more effective than vitamin E [J]. Life Sciences, 55: 271-276.

PORAT R, LICHTER A, TERRY L A, et al., 2018. Postharvest losses of fruit and vegetables during retail and in consumers' homes: quantifications, causes, and means of prevention [J]. Postharvest Biology and Technology, 139: 135-149.

POSE S, PANIAGUA C, MATAS A J, et al., 2019. A nanostructural view of the cell wall disassembly process during fruit ripening and postharvest storage by atomic force microscopy [J]. Trends in Food Science & Technology, 87: 47-58.

POSMYK M M, JANAS K M, 2009. Melatonin in plants [J]. Acta Physiologiae Plantarum, 31: 1-11.

PRUTHI J S, 1963. Physiology, chemistry, and technology of passion fruit [J]. Advances in Food Research, 12: 203-282.

PURGATTO E, LAJOLO F M, NASCIMENTO J R, et al., 2001. Inhibition of β-amylase activity, starch degradation and sucrose formation by indole-3-acetic acid during banana ripening [J]. Planta, 212: 823-828.

QI X L, DONG Y X, LIU C L, et al., 2022. The PavNAC56 transcription factor positively regulates fruit ripening and softening in sweet cherry (*Prunus avium*) [J]. Physiologia Plantarum, 174: e13834.

QI X, LIU C, SONG L, et al., 2020. PaMADS7, a MADS-box transcription factor, regulates sweet cherry fruit ripening and softening [J]. Plant Science, 301: 110634.

QIAN C L, JI Z J, ZHU Q, et al., 2021. Effects of 1-MCP on proline, polyamine, and nitric oxide metabolism in postharvest peach fruit under chilling stress [J]. Horticultural Plant Journal, 7: 188-196.

QIAN M, XU Z, ZHANG Z, et al., 2021. The downregulation of PpPG21 and PpPG22 influences peach fruit texture and softening [J]. Planta, 254 (2): 22.

QIN G, WANG Y, CAO B, et al., 2012. Unraveling the regulatory network of the MADS box transcription factor RIN in fruit ripening [J]. Plant Journal, 70 (2): 243-255.

QIN W, PAPPAN K, WANG X, 1997. Molecular heterogeneity of phospholipase D (PLD). Cloning of PLD gamma and regulation of plant PLD gamma, -beta, and -alpha by polyphosphoinositides and calcium [J]. Journal of Biology Chemistry, 272: 28267-28273.

RABIEI V, KAKAVAND F, ZAAREKMAHANDI F, et al., 2019. Nitric oxide and γ-aminobutyric acid treatments delay senescence of cornelian cherry fruits during postharvest cold storage by enhancing antioxidant system activity-sciencedirect [J]. Scientia Horticulturae, 243: 268-273.

RAFFO A, KELDERER M, PAOLETTI F, et al., 2009. Impact of innovative controlled atmosphere storage technologies and postharvest treatments on volatile compound production in cv. Pinova apples [J]. Journal of Agricultural and Food Chemistry, 57 (3): 915-923.

RAZAVI F, MAHMOUDI R, RABIEI V, et al., 2018. Glycine betaine treatment attenuates chilling injury and maintains nutritional quality of hawthorn fruit during storage at low temperature [J]. Scientia Horticulturae, 233: 188-194.

RAZZAQ K, KHAN A S, MALIK A U, et al., 2014. Role of putrescine in regulating fruit softening and antioxidative enzyme systems in 'Samar Bahisht Chaunsa' mango [J]. Postharvest Biology and Technology, 96: 23-32.

REN Y, YAN T, HU C, et al., 2023. Exogenous Nitric OxideInduced Postharvest Gray Spot Disease Resistance in Loquat Fruit and Its Possible Mechanism of Action [J]. International Journal of Molecular Sciences, 24: 4369.

RIBEIRO J S, VELOSO C M, 2021. Microencapsulation of natural dyes with biopolymers for application in food: a review [J]. Food Hydrocolloids, 112: 106374.

RIECHMANN J L, HEARD J, MARTIN G, et al., 2000. Arabidopsis transcription factors: genome-wide comparative analysis among eukaryotes [J]. Science, 290 (5499): 2105-2110.

ROBERTSON J A, MEREDITH F I, HORVAT R J, et al., 1990. Effect of cold storage and maturity on the physical and chemical characteristics and volatile constituents of peaches (Cv. Cresthaven) [J]. Journal of Agricultural and Food Chemistry, 38 (3): 53-70.

RODRIGUEZ C, MAYO J C, SAINZ R M, et al., 2004. Regulation of antioxidant enzymes: A significant role for melatonin [J]. Journal of Pineal Research, 36: 1-9.

RODRIGUEZ-AMAYA D B, 2019. Update on natural food pigments-a mini-review on carotenoids, anthocyanins, and betalains [J]. Food Research International, 124: 200-205.

ROSE J K C, BENNETT A B, 1999. Cooperative disassembly of the cellulose-xyloglucan network of plant cell walls: Parallels between cell expansion and fruit ripening [J]. Trends in Plant Science, 4: 176-183.

SABATO D, ESTERAS C, GRILLO O, et al., 2015. Seeds morpho-colourimetric analysis as complementary method to molecular characterization of melon diversity [J]. Scientia Horticulturae, 192: 441-452.

SAHAY S, GUPTA M, 2017. An update on nitric oxide and its benign role in plant responses under metal stress [J]. Nitric Oxide, 67: 39-52.

SANG Y Y, YANG W T, LIU Y X, et al., 2022. Influences of low temperature on the postharvest quality and antioxidant capacity of winter jujube (Zizyphus jujuba Mill. cv. Dongzao) [J]. LWT-Food Science and Technology, 154: 112876.

SANG Y, YANG W, LIU Y, et al., 2021. Influences of low temperature on the postharvest quality and antioxidant capacity of winter jujube (*Zizyphus jujuba* mill. cv. dongzao) -sciencedirect [J]. LWT-Food Science and Technology, 154: 1-9.

SCOTTICAMPOS P, PAIS I P, PARTELLI F L, et al., 2014. Phospholipids profile in chloroplasts of coffea spp. genotypes differing in cold acclimation ability [J]. Journal of Plant Physiology, 171: 243-249.

SEVILLANO L, SANCHEZ-BALLESTA M T, ROMOJARO F, et al., 2009. Physiological, hormonal and molecular mechanisms regulating chilling injury in horticultural species. Postharvest technologies applied to reduce its impact [J]. Journal of the Science of Food and Agriculture, 89: 555-573.

SEYMOUR G B, ØSTERGAARD L, CHAPMAN N H, et al., 2013. fruit development and ripening [J]. Annual Review of Plant Biology, 64: 219-241.

SHAN T M, JIN P, ZHANG Y, et al., 2016. Exogenous glycine betaine treatment enhances chilling tolerance of peach fruit during cold storage [J]. Postharvest Biology and Technology, 114: 104-110.

SHAN W, GUO Y F, WEI W, et al., 2020. Banana MaBZR1/2 associate with MaMPK14 to modulate cell wall modifying genes during fruit ripening [J]. Plant Cell Reports, 39: 35-46.

SHAN Y, ZHANG D, LUO Z, et al., 2022. Advances in chilling injury of postharvest fruit and vegetable: Extracellular ATP aspects [J]. Comprehensive Reviews in Food Science and Food Safety, 21: 4251-4273.

SHANG H, CAO S, YANG Z, et al., 2011. Effect of exogenous γ-aminobutyric acid treatment on proline accumulation and chilling injury in peach fruit after long-term cold storage [J]. Journal of Agricultural & Food Chemistry, 59 (4): 1264-1268.

SHARMA S, KRISHNA H, BARMAN K, et al., 2022. Synergistic effect of polyamine treatment and chitosan coating on postharvest senescence and enzyme activity of bell pepper (*Capsicum annuum* L.) fruit [J]. South African Journal of Botany, 151: 175-184.

SHARMA S, SHARMA R R, 2015. Nitric oxide inhibits activities of pal and pme enzymes and reduces chilling injury in 'santa rosa' japanese plum (prunus salicina lindell) [J]. Journal of Plant Biochemistry & Biotechnology, 24 (3): 292-297.

SHARMA S, SHARMA R R, VERMA M K, 2015. Postharvest treatment with nitric oxide influences the physiological and quality attributes of 'santa rosa' plums during cold storage [J]. Indian Journal of Horticulture, 72 (4), 535-540.

SHARMA S, SHARMA R, PAL R K, et al., 2012. Ethylene absorbents influence fruit firmness and activity of enzymes involved in fruit softening of Japanese plum (Prunus salicina Lindell.) cv. Santa Rosa [J]. Fruits, 67 (4): 257-266.

SHARMA S, SINGH A K, SINGH S K, et al., 2019. Polyamines for preserving postharvest quality. Emerging Postharvest Treatment of Fruits and Vegetables [M]. USA: Apple Academic Press.

SHELP B J, BOWN A W, ZAREI A, 2017. 4-Aminobutyrate (GABA): A metabolite and signal with practical significance [J]. Botany, 95 (11): 1015-1032.

SHEN Y H, ZHAO Z L, ZHANG L Y, et al., 2017. Metabolic activity induces membrane phase separation in endoplasmic reticulum [J]. Proceedings of the National Academy of Sciences, 114: 1712555114.

SHENG L, SHEN D D, LUO Y, et al., 2017. Exogenous γ-aminobutyric acid treatment affects citrate and amino acid accumulation to improve fruit quality and storage performance of postharvest citrus fruit [J]. Food Chemistry, 216: 138-145.

SHENG L, ZHOU X, LIU Z Y, et al., 2016. Effect of low temperature storage on energy and lipid metabolisms accompanying peel browning of 'Nanguo' pears during shelf life [J]. Postharvest Biology and Technology, 117: 1-8.

SHENG S, LI T, HAI L R, 2018. Corn phytochemicals and their health benefits [J]. Food ence and Human Wellness, 7: 185-195.

SHI F, ZHOU X, YAO M M, et al., 2019. Low-temperature stressinduced aroma loss by regulating fatty acid metabolism pathway in 'Nanguo' pear [J]. Food Chemistry, 297: 124927.

SHI S H D, LIU Z C, WANG J W, et al., 2019. Nitric oxide modulates sugar metabolism and maintains the quality of red raspberry during storage [J]. Scientia Horticulturae, 256: 108611.

SHI Y, LI B J, SU G, et al., 2022. Transcriptional regulation of fleshy fruit texture [J]. Journal of Integrative Plant Biology, 64: 1649-1672.

SHIGA T M, SOARES C A, NASCIMENTO J R, et al., 2011. Ripening-associated changes in the amounts of starch and non-starch polysaccharides and their contributions to fruit softening in three banana cultivars [J]. J. Sci. Food Agric., 91: 1511-15166.

SHRESTHA R, GOMEZ-ARIZA J, BRAMBILLA V, et al., 2014. Molecular control of seasonal flowering in rice, Arabidopsis and temperate cereals [J]. Annals of Botany, 114 (7): 1445-1458.

SILVA D C, FREITAS A L P, BARROS F C N, et al., 2012. Polysaccharide isolated from passiflora edulis: characterization and antitumor properties [J]. Carbohydrate Polymers, 87 (1): 139-145.

SILVA J K D, CAZARIN C B B, COLOMEU T C, et al., 2013. Antioxidant activity of aqueous extract of passion fruit (passiflora edulis) leaves: in vitro and in vivo study [J]. Food Research International, 53 (2): 882-890.

SILVIA E L, MARÍA L M, JOSE L B, et al., 2019. Calcium chloride treatment modifies cell wall metabolism and activates defense responses in strawberry fruit (Fragaria × ananassa, Duch) [J]. Journal of the Science of Food and Agriculture, 99 (8): 4003-4010.

SINGH V, JAWANDHA S K, GILL P P S, 2020. Putrescine application reduces softening and maintains the quality of pear fruit during cold storage [J]. Acta Physiologiae Plantarum, 42: 932-936.

SINHA A, GILL P P S, JAWANDHA S K, et al., 2022. Composite coating of chitosan with salicylic acid retards pear fruit softening under cold and supermarket storage [J]. Food Research International, 160: 11724.

SIRIKESORN L, KETSA S, VAN DOORN W G, 2013. Low temperature-induced water-soaking of Dendrobium inflorescences: Relation with phospholipase D activity, thiobarbaturic-acid-staining membrane degradation products, and membrane fatty acid composition [J]. Postharvest Biology and Technology, 80: 47-55.

SIVAKUMAR D, TERRY L A, KORSTEN L, 2010. An overview on litchi fruit quality and alternative postharvest treatments to replace sulfur dioxide fumigation [J]. Food Reviews International, 26 (2): 162-188.

SIYUAN S, TONG L, LIU R H, 2018. Corn phytochemicals and their health benefits [J]. Food Science and Human Wellness, 7 (3): 185-195.

SLASKI J J, ZHANG G C, BASU U, et al., 1996. Aluminum resistance in wheat (*Triticum aestivum*) is associated with rapid, al-induced changes in activities of glucose-6-phosphate dehydrogenase and 6-phosphogluconate dehydrogenase in root apices [J]. Physiologia Plantarum, 98 (3): 477-484.

SMITH D L, ABBOTT J A, GROSS K C, 2002. Down-regulation of tomato β-galactosidase 4 results in decreased fruit softening [J]. Plant Physiology, 129 (4): 1755-1762.

SONG H M, XU X B, WANG H, et al., 2010. Exogenous γ-aminobutyric acid alleviates oxidative damage caused by aluminium and proton stresses on barley seedlings [J]. Journal of the Science of Food and Agriculture, 90: 1410-1416.

SONG J, BANGETH F, 2003. Fatty acids as precursors for aroma volatile biosynthesis in pre-climacteric and climacteric apple fruit [J]. Postharvest Biology and Technology, 30 (2): 113-121.

SONG X DAI H, WANG S, et al., 2022. Putrescine Treatment Delayed the Softening of Postharvest Blueberry Fruit by Inhibiting the Expression of Cell Wall Metabolism Key Gene VcPG1 [J]. Plants, 11: 1356.

SONG Z Y, QIAO J, TIAN D D, et al., 2023. Glutamic acid can prevent the browning of fresh-cut potatoes by inhibiting PPO activity and regulating amino acid metabolism [J]. LWT - Food Science and Technology, 180: 114735.

STRACKE R, WERBER M, WEISSHAAR B, 2001. The R2R3-MYB gene family in Arabidopsis thaliana [J]. Current Opinion in Plant Biology, 4: 447-456.

SUN B, LIN P X, XIA P X, et al., 2020. Low-temperature storage after harvest retards the deterioration in the sensory quality, health-promoting compounds, and antioxidant capacity of baby mustard [J]. Royal Society of Chemistry Advances, 10 (60): 36495-36503.

SUN D, LIANG G, XIE J, et al., 2010. Improved preservation effects of lychee fruit by combining chitosan coating with ascorbic acid treatment during postharvest storage [J]. Afrian Journal of Biotechnology, 9: 3272-3279.

SUN H J, LUO M L, ZHOU X, et al., 2020. Exogenous glycine betaine treatment alleviates low temperature-induced pericarp browning of 'Nanguo' pears by regulating antioxidant enzymes and proline metabolism [J]. Food Chemistry, 306: 125626.

SUN J, CHU Y F, WU X, LIU R H, 2002. Antioxidant and antiproliferative activities of common fruits. Journal of Agricultural and Food Chemistry, 50: 7449-7454.

SUN Q, ZHANG N, WANG J, et al., 2015. Melatonin promotes ripening and improves quality of tomato fruit during postharvest life [J]. Journal of Experimental Botany, 66: 657-668.

SUN X, ZHANG L, CAO Y, et al., 2016. Quantitative Analysis and Comparison of Four Major Flavonol Glycosides in the Leaves of Toona sinensis (A. Juss.) Roemer (Chinese Toon) from Various Origins by High-Performance Liquid Chromatography-Diode Array Detector and Hierarchical Clustering Analysis [J]. Pharmacognosy Magazine, 12 (12): 270-276.

SUTHERLAND P W, FULLERTON C G, SCHRÖDER R, et al., 2021. Cell wall changes in Actinidia arguta during softening [J]. Scientia Horticulturae, 226: 173-183.

TACKEN E, IRELAND H, GUNASEELAN K, et al., 2010. The role of ethylene and cold temperature in the regulation of the apple polygalacturonase1 gene and fruit softening [J]. Plant Physiology, 153 (1): 294-305.

TAKESHIMA K, YAMATSU A, YAMASHITA Y, et al., 2014. Subchronic toxicity evaluation of γ-aminobutyric acid (GABA) in rats [J]. Food & Chemical Toxicology, 68: 128-134.

TAN D X, CHEN L D, POEGGELER B, et al., 1993. Melatonin: a potent, endogenous hydroxyl radical scavenger [J]. Endocrine Journal, 1: 57-60.

TAN X L, FAN Z Q, ZENG Z X, et al., 2021. Exogenous melatonin maintains leaf quality of postharvest chinese flowering cabbage by modulating respiratory metabolism and energy status [J]. Postharvest Biology and Technology, 177: 111524.

TANG Y, ZHANG C, CAO S, et al., 2015. The effect of CmLOXs on the production of volatile organic compounds in four aroma types of melon (Cucumis melo). PLoS One, 10: e0143567.

TEIXEIRA G H D A, SANTOS L O, LUIS C C, et al., 2017. Effect of carbon dioxide (CO_2) and oxygen (O_2) levels on quality of 'palmer' mangoes under controlled atmosphere storage [J]. Journal of Food Science and Technology, 55 (1): 145-156.

TESZLÁK P, GAÁL K, NIKFARDJAM M S P, 2005. Influence of grapevine flower treatment with gibberellic acid (GA_3) on polyphenol content of *Vitis vinifera* L. wine [J]. Analytica Chimica Acta, 543 (1-2): 275-281.

THEIßEN G, MELZER R, RÜMPLER F, 2016. MADS-domain transcription factors and the floral quartet model of flower development: linking plant development and evolution [J]. Development, 143: 3259-3271.

THEWES F R, BRACKMANN A, ANESE R, et al., 2018. 1-methylcyclopropene suppresses anaerobic metabolism in apples stored under dynamic controlled atmosphere monitored by respiratory quotient [J]. Scientia Horticulturae, 227 (1): 288.

THEWES F R, BRACKMANN A, & A, N. D, 2019. Dynamics of sugars, anaerobic metabolism enzymes and metabolites in apples stored under dynamic controlled atmosphere [J]. Scientia Horticulturae, 255: 145-152.

TIAN S P, QIN G, LI B, 2013. Reactive oxygen species involved in regulating fruit senescence and fungal pathogenicity [J]. Plant Molecular Biology, 82 (6): 593-602.

TIAN X, ZHU L, YANG N, et al., 2021. Proteomics and Metabolomics Reveal the Regulatory Pathways of Ripening and Quality in Post-Harvest Kiwifruits [J]. Journal of Agricultural and Food Chemistry, 69 (2): 824-835.

TIEMAN, D, 2017. Transcriptional control of strawberry ripening-two to tango [J]. Journal of Experimental Botany, 68 (16): 4407-4409.

TIETEL Z, PORAT R, WEISS K, et al., 2011. Identification of aroma-active compounds in fresh and stored 'Mor' mandarins [J]. International Journal of Food Science & Technology, 46 (11): 2225-2231.

TOURNIER B, SANCHEZ M T, JONES B, et al., 2003. New members of the tomato ERF family show specific expression pattern and diverse DNA-binding capacity to the GCC box element [J]. Febs Letters, 550 (1-3): 149-154.

TRAN T T L, AIAMLA-OR S, SRILAONG V, et al., 2015. Application of nitric oxide to extend the shelf life of mango fruit [J]. Acta Horticulturae (1088): 97-102.

TUCKER G, YIN X R, ZHANG A D, et al., 2017. Ethylene and fruit softening [J]. Food Quality and Safety, 1: 253-267.

TURMANIDZE T, JGENTI M, GULUA L, et al., 2017. Effect of ascorbic acid treatment on some quality parameters of frozen strawberry and raspberry fruits [J]. Annals of Agrarian Science, 15 (3): 370-374.

TURNBULL J J, NAKAJIMA J, WELFORD R W, et al., 2004. Mechanistic studies on three 2-oxoglutarate-dependent oxygenases of flavonoid biosynthesis: Anthocyanidin synthase, flavonol synthase, and flavanone 3beta-hydroxylase [J]. Journal of Biological Chemistry, 279 (2): 1206-1216.

UEKI J, MORIOKA S, KOMARI T, et al., 1995. Purification and characterization of phospholipase D (PLD) from (*Oryza sativa* L.) and cloning of cDNA for PLD from rice and maize (*Zea mays* L.) [J]. Plant Cell Physiology, 36: 903-914.

VALENZUELA J L, MANZANO S, PALMA F, et al., 2017. Oxidative stress associated with chilling injury in immature fruit: postharvest technological and biotechnological solutions [J]. International Journal of Molecular Sciences, 18 (7): 1467.

VEIGA J C, SILVEIRA N M, SEABRA A B, et al., 2024. Exploring the power of nitric oxide and nanotechnology for prolonging postharvest shelf-life and enhancing fruit quality [J]. Nitric Oxide, 142: 26-37.

VELTMAN R H, LARRIGAUDIERE C, WICHERS H J, et al., 1999. PPO activity and polyphenol content are not limiting factors during brown core development in pears (*Pyrus communis* L. cv. Conference) [J]. Journal of Plant Physiology, 154: 697-702.

VELTMAN R H, LENTHÉRIC I, PLAS L H W V D, et al., 2003. Internal browning in pear fruit (Pyrus communis L. cv Conference) may be a 20 result of a limited availability of energy and antioxidants [J]. Postharvest Biology & Technology, 28 (2): 295-302.

VENDRELL M, 1969. Reversion of senescence-Effects of 2, 4-dichlorophenoxyacetic acid and indoleacetic acid on respiration ethylene production and ripening of banana fruit slices [J]. Australian Journal of Biological Sciences, 22: 601-610.

VICHAIYA T, FAIYUE B, ROTARAYANONT S, et al., 2022. Exogenous trehalose alleviates chilling injury of 'Kim Ju' guava by modulating soluble sugar and energy metabolisms [J]. Scientia Horticulturae, 301: 11138.

VILLARREAL N M, BUSTAMANTE C A, CIVELLO P M, et al., 2010. Effect of ethylene and 1-MCP treatments on strawberry fruit ripening [J].

Journal of the science of food and agriculture, 90: 683-689.

VILLATORO, C, ECHEVERRÍA G, GRAELL J, et al., 2008. Long-term storage of Pink Lady apples modifies volatile-involved enzyme activities: consequences on production of volatile esters [J]. Journal of Agriculture and Food Chemistry, 56: 9166-9174.

VREBALOV J, RUEZINSKY D, PADMANABHAN V, et al., 2002. A MADS-box gene necessary for fruit ripening at the tomato ripening-inhibitor (rin) locus [J]. Science, 296 (5566): 343-346.

WANG A Y, ZHOU F H, ZUO J H, et al., 2018. Pre-storage treatment of mechanically-injured green pepper (*Capsicum annuum* L.) fruit with putrescine reduces adverse physiological responses [J]. Postharvest Biology and Technology, 145: 239-246.

WANG D D, SEYMOUR G B, 2022. Molecular and biochemical basis of oftening in tomato [J]. Molecular Horticulture. 2: 1-10.

WANG D D, YEATS T H, ULUISIK S, et al., 2018. Fruit Softening: Revisiting the Role of Pectin [J]. Trends Plant Science, 23: 302-310.

WANG F, ZHANG X, YANG Q, et al., 2019b. Exogenous melatonin delays postharvest fruit senescence and maintains the quality of sweet cherries [J]. Food Chemistry, 301: 125311.

WANG H, CHEN Y, LIN H, et al., 2020. 1-Methylcyclopropene containing-papers suppress the disassembly of cell wall polysaccharides in Anxi persimmon fruit during storage [J]. International Journal of Biological Macromolecules, 151: 723-729.

WANG H, QIAN Z, MA S, et al., 2013. Energy status of ripening and postharvest senescent fruit of litchi (*Litchi chinensis* Sonn.) [J]. BMC Plant Biology, 13 (1): 1-16.

WANG J W, LV M, LI G D, et al., 2018b. Effect of intermittent warming on alleviation of peel browning of 'Nanguo' pears by regulation energy and lipid metabolisms after cold storage [J]. Postharvest Biology and Technology, 142: 99-106.

WANG J W, ZHOU X, ZHOU Q, et al., 2017a. Low temperature conditioning alleviates peel browning by modulating energy and lipid metabolisms of 'Nanguo' pears during shelf life after cold storage [J]. Postharvest

Biology and Technology, 131: 10-15.

WANG J, LV M, HE H, et al., 2020. Glycine betaine alleviated peel browning in cold-stored 'Nanguo' pears during shelf life by regulating phenylpropanoid and soluble sugar metabolisms [J]. Scientia Horticulturae, 262: 109100.

WANG K, SHAO X F, GONG Y F, et al., 2013. The metabolism of soluble carbohydrates related to chilling injury in peach fruit exposed to cold stress [J]. Postharvest Biology and Technology, 86: 53-61.

WANG L, BOKHARY S U F, XIE B, et al., 2019a. Biochemical and molecular effects of glycine betaine treatment on membrane fatty acid metabolism in cold stored peaches [J]. Postharvest Biology and Technology, 154: 58-69.

WANG L, BOKHARY S U F, XIE B, et al., 2019b. Biochemical and molecular effects of glycine betaine treatment on membrane fatty acid metabolism in cold stored peaches [J]. Postharvest Biology and Technology, 154: 58-69.

WANG L, FENG C, ZHENG X, et al., 2017. Plant mitochondria synthesize melatonin and enhance the tolerance of plants to drought stress [J]. Journal of Pineal Research, 63 (3): 12429.

WANG L, SHAN T M, XIE B, et al., 2019. Glycine betaine reduces chilling injury in peach fruit by enhancing phenolic and sugar metabolisms [J]. Food Chemistry, 272: 530-538.

WANG L, SHI Y, WANG R, et al., 2021. Antioxidant activity and healthy benefits of natural pigments in fruits: a review [J]. International Journal of Molecular Sciences, 22 (9): 4945.

WANG N, NIAN Y, LI R, et al., 2022. Transcription factor *CpbHLH3* and *CpXYN1* gene cooperatively regulate fruit texture and counteract 1-methylcyclopropene inhibition of softening in postharvest papaya (*Carica papaya* L.) [J]. Journal of Agricultural and Food Chemistry, 70 (32): 9919-9930.

WANG Q, DING T, ZUO J H, et al., 2016. Amelioration of postharvest chilling injury in sweet pepper by glycine betaine [J]. Postharvest Biology and Technology, 112: 114-120.

WANG X, 2002. Phospholipase D in hormonal and stress signaling [J]. Current Opinion in Plant Biology, 5: 408-414.

WANG X, 2005. Regulatory functions of phospholipase D and phosphatidic acid in plant growth, development, and stress responses [J]. Plant Physiology, 139: 566-573.

WANG X, PAN L, WANG Y, et al., 2021. PpIAA1 and PpERF4 form a positive feedback loop to regulate peach fruit ripening by integrating auxin and ethylene signals [J]. Plant Science, 313: 111084.

WANG X, XU L, ZHENG L, 1994. Cloning and expression of phosphatidylcholine-hydrolyzing phospholipase D from Ricinus communis L [J]. Journal of Biology Chemistry, 269: 20312-20317.

WANG Y H, YAN Z M, TANG W H, et al., 2021. Impact of chitosan, sucrose, glucose, and fructose on the postharvest decay, quality, enzyme activity, and defense-related gene expression of strawberries [J]. Horticulturae, 7 (12): 518.

WANG Y, LUO Z, HUANG X, et al., 2014. Effect of exogenous γ-aminobutyric acid (GABA) treatment on chilling injury and antioxidant capacity in banana peel [J]. Scientia Horticulturae, 168: 132-137.

WANG Y, XIE X, LONG L E, 2014. The effect of postharvest calcium application in hydro-cooling water on tissue calcium content, biochemical changes, and quality attributes of sweet cherry fruit [J]. Food Chemistry, 160: 22-30.

WANG Y, XU F, FENG X, et al., 2015. Modulation of Actinidia arguta fruit ripening by three ethylene biosynthesis inhibitors [J]. Food Chemistry, 173: 405-413.

WANG Z, CAO J K, JIANG W B, 2016. Changes in sugar metabolism caused by exogenous oxalic acid related to chilling tolerance of apricot fruit [J]. Postharvest Biology & Technology, 114 (9): 10-16.

WEBER A, THEWES F R, SELLWIG M, et al., 2019. Dynamic controlled atmosphere: Impact of elevated storage temperature on anaerobic metabolism and quality of 'Nicoter' apples [J]. Food Chemistry, 298: 125017.

WEI J, MIAO H Y, WANG Q M, 2011. Effect of glucose on glucosinolates, antioxidants and metabolic enzymes in Brassica sprouts [J]. Scientia Horticulturae, 129 (4): 535-540.

WEI S, QIN G, ZHANG H, et al., 2017. Calcium treatments promote the aroma volatiles emission of pear (Pyrus ussuriensis 'Nanguoli') fruit during post-harvest ripening process [J]. Scientia Horticulturae, 215: 102-111.

WELFORD R W, CLIFTON I J, TURNBULL J J, et al., 2005. Structural and mechanistic studies on anthocyanidin synthase catalyzed oxidation of flavanone substrates: the effect of C-2 stereochemistry on product selectivity and mechanism [J]. Organic and Biomolecular Chemistry, 3 (17): 3117-3126.

WELTI R, LI W, LI M, et al., 2002. Profiling membrane lipids in plant stress responses [J]. Journal of Biology Chemistry, 277: 31994-32002.

WESTON K, 1998. MYB proteins in life, death and differentiation [J]. Current Opinion in Genetics and Development, 8 (1): 76-81.

WHITAKER B D, SMITH D L, GREEN K C, 2001. Cloning, characterization and functional expression of a phospholipase Dalpha cDNA from tomato fruit [J]. Physiology Plant, 112: 87-94.

WILKINSON J Q, LANAHAN M B, YEN H C, et al., 1995. An ethylene-inducible component of signal transduction encoded by never-ripe [J]. Science, 270 (5243): 1807-1809.

WILLETT W, ROCKSTRÖM J, LOKEN B, et al., 2019. Food in the Anthropocene: the EAT-Lancet Commission on healthy diets from sustainable food systems [J]. Lancet, 393: 447-492.

WILLS R B H, KU V V V, LESHEM Y Y, 2000. Fumigation with nitric oxide to extend the postharvest life of strawberries [J]. Postharvest Biology & Technology, 18 (1): 75-79.

WILLS R B H, SOEGIARTO L, BOWYER M C, 2007. Use of a solid mixture containing diethylenetriamine/nitric oxide (detano) to liberate nitric oxide gas in the presence of horticultural produce to extend postharvest life [J]. Nitric Oxide Biology & Chemistry, 17 (1): 44-49.

WIN N M, YOO J, KWON S I, et al., 2019. Characterization of fruit quality attributes and cell wall metabolism in 1-methylcyclopropene (1-MCP) -treated 'Summer King' and 'Green Ball' apples during cold storage [J].

Frontiers in Plant Science, 10: 1513.

WU W, WANG M M, GONG H, et al., 2020. High CO_2/hypoxia-induced softening of persimmon fruit is modulated by DkERF8/16 and DkNAC9 complexes [J]. Journal of Experimental Botany, 71: 2690-2700.

XIAO Y Y, CHEN J Y, KUANG J F, et al., 2013. Banana ethylene response factors are involved in fruit ripening through their interactions with ethylene biosynthesis genes [J]. Journal of Experimental Botany, 64 (8): 2499-2510.

XING C H, LIU Y, ZHAO L Y, et al., 2019. A novel MYB transcription factor regulates AsA synthesis and effect cold tolerance [J]. Plant Cell and Environment, 42: 832-845.

XING W, RAJASHEKAR C B, 2001. Glycine betaine involvement in freezing tolerance and water stress in arabidopsis thaliana [J]. Environmental and Experimental Botany, 46 (1): 21-28.

XU D, ZHOU F, GU S, et al., 2021. 1-Methylcyclopropene maintains the postharvest quality of hardy kiwifruit (*Actinidia aruguta*) [J]. Journal of Food Measurement and Characterization, 15: 3036-3044.

XU M J, DONG J F, ZHANG M, et al., 2012. Cold-induced endogenous nitric oxide generation plays a role in chilling tolerance of loquat fruit during postharvest storage [J]. Postharvest Biology & Technology, 65: 5-12.

YAHIA E M, FADANELLI L, MATTÈ P, et al., 2019. Controlled atmosphere storage // Postharvest technology of perishable horticultural commodities [M]. Cambridge: Woodhead Publishing.

YAMASAKI K, KIGAWA T, INOUE M, et al., 2005. Solution structure of an Arabidopsis WRKY DNA binding domain [J]. The Plant Cell, 17 (3): 944-956.

YAN F, CAI T, WU Y Y, et al., 2021. Physiological and transcriptomics analysis of the effect of recombinant serine protease on the preservation of loquat [J]. Genomics, 113 (6): 3750-3761.

YAN Y, LI Z, MATTHEOS A G, 2008. Koffas. High-yield anthocyanin biosynthesis in engineered Escherichia coli [J]. Biotechnology and Bioengineering, 100 (1): 126-140.

YANG A, CAO S, YANG Z, et al., 2011. γ-Aminobutyric acid treatment reduces chilling injury and activates the defence response of peach fruit [J]. Food Chemistry, 129 (4): 1619-1622.

YANG C, LU X, MA B, et al., 2015. Ethylene signaling in rice and Arabidopsis: Conserved and diverged aspects [J]. Molecular Plant, 8 (4): 495-505.

YANG H, WU F, CHENG J, 2011. Reduced chilling injury in cucumber by nitric oxide and the antioxidant response [J]. Food Chemistry, 127 (3): 1237-1242.

YANG J, SUN C, FU D, et al., 2017. L-glutamate inhibition of growth of Alternaria alternata by inducing resistance in tomato fruit [J]. Food Chemistry, 230: 145-153.

YAO W, XU T, FAROOQ S U, et al., 2018. Glycine betaine treatment alleviates chilling injury in zucchini fruit (*cucurbita pepo*, L.) by modulating antioxidant enzymes and membrane fatty acid metabolism [J]. Postharvest Biology and Technology, 144: 20-28.

YAZAKI K, 2005. Transporters of secondary metabolites [J]. Current Opinion in Plant Biology, 8: 301-307.

YI C, QU H X, JIANG Y M, et al., 2008. ATP induced changes in energy status and membrane integrity of harvested litchi fruit and its relation to pathogen resistance [J]. Journal of Phytopathology, 156: 365-371.

YI J W, WANG YI, MA X S, et al., 2021. LcERF2 modulates cell wall metabolism by directly targeting a UDP-glucose-4-epimerase gene to regulate pedicel development and fruit abscission of litchi [J]. Plant Journal, 106 (3): 801-816.

YI X K, LIU G F, RANA M M, et al., 2016. Volatile profiling of two pear genotypes with different potential for white pear aroma improvement [J]. Scientia Horticulturae, 209: 221-228.

YIN X R, ALLAN A C, CHEN K S, et al., 2010. Kiwifruit EIL and ERF genes involved in regulating fruit ripening [J]. Plant Physiology, 153 (3): 1280-1292.

YOKOYAMA R, SHINOHARA N, ASAOKA R, et al., 2014. The biosynthesis and function of polysaccharide components of the plant cell wall [J]. In Plant Cell Wall Patterning and Cell Shape, 1: 3-34.

YUAN J, MISHRA P, CHING C B, 2017. Engineering the leucine biosynthetic pathway for isoamyl alcohol overproduction in Saccharomyces cerevisiae [J]. Journal of Industrial Microbiology and Biotechnology, 44: 107-117.

YUAN X Y, WANG R H, ZHAO X D, et al., 2016. Role of the tomato non-ripening mutation in regulating fruit quality elucidated using iTRAQ Protein profile analysis [J]. PLoS One, 11 (10): e0164335.

ZAHARAH S S, SINGH Z, 2011. Postharvest nitric oxide fumigation alleviates chilling injury, delays fruit ripening and maintains quality in cold-stored 'kensington pride' mango [J]. Postharvest Biology & Technology, 60 (3): 202-210.

ZAHEDI S M, HOSSEINI M S, KARIMI M, et al., 2019. Effects of postharvest polyamine application and edible coating on maintaining quality of mango (*Mangifera indica* L.) cv. Langra during cold storage [J]. Food Science & Nutrition, 7: 433-441.

ZENG J K, CHEN C Y, CHEN M, et al., 2022. Comparative transcriptomic and metabolomic analyses reveal the delaying effect of naringin on postharvest decay in citrus fruit [J]. Frontiers in Plant Science, 13: 1045857.

ZHAI Z F, FENG C, WANG Y Y, et al., 2021. Genome-Wide Identification of the Xyloglucan endotransglucosylase/Hydrolase (XTH) and Polygalacturonase (PG) Genes and Characterization of Their Role in Fruit Softening of Sweet Cherry [J]. International Journal of Molecular Sciences, 22: 12331.

ZHANG B, GAO Y, ZHANG L, et al., 2021. The plant cell wall: Biosynthesis, construction, and functions [J]. Journal of Integrative Plant Biology, 63 (1): 251-272.

ZHANG C F, TIAN S P, 2010. Peach fruit acquired tolerance to low temperature stress by accumulation of linolenic acid and N-acylphosphatidylethanolamine in plasma membrane [J]. Food Chemistry, 120: 864-872.

ZHANG H, WANG R, WANG T, et al., 2019. Methyl salicylate delays peel yellowing of 'aosu' pear (Pyrus bretschneideri) during storage by regulating chlorophyll metabolism and maintaining chloroplast ultrastructure [J]. Journal of the Science of Food and Agriculture, 99: 4816-4824.

ZHANG J Y, WANG C, CHEN C K, et al., 2023. Glycine betaine inhibits postharvest softening and quality decline of winter jujube fruit by regulating energy and antioxidant metabolism [J]. Food Chemistry, 410: 135445.

ZHANG J, KIRKHAM M B, 1996. Lipid peroxidation in sorghum and sunflower seedlings as affected by ascorbic acid, benzoic acid, and propyl gallate [J]. Journal of Plant Physiology, 149 (5): 489-493.

ZHANG J, LIU S, ZHU X M, et al., 2023. A comprehensive evaluation of tomato fruit quality and identification of volatile compounds [J]. Plants-Basel, 12 (16): 2947.

ZHANG J, WEN M, DAI R, et al., 2023. Comparative physiological and transcriptome analyses reveal mechanisms of salicylic - acid - reduced postharvest ripening in 'hosui' pears (*Pyrus pyrifolia* Nakai) [J]. Plants, 12: 3429.

ZHANG L L, WANG X F, DONG K, et al., 2024. Tandem transcription factors PpNAC1 and PpNAC5 synergistically activate the transcription of the PpPGF to regulate peach softening during fruit ripening [J]. Plant Molecular Biology, 114: 46.

ZHANG L, WANG P, CHEN F, et al., 2019. Effects of calcium and pectin methylesterase on quality attributes and pectin morphology of jujube fruit under vacuum impregnation during storage [J]. Food Chemistry, 289 (15): 40-48.

ZHANG Q Y, GE J, LIU X C, et al., 2022. Consensus co - expression network analysis identifies AdZAT5 regulating pectin degradation in ripening kiwifruit [J]. Journal of Advanced Research, 40: 59-68.

ZHANG S, LIN Y, LIN H, et al., 2018. Lasiodiplodia theobromae (pat.) griff. & maubl. -induced disease development and pericarp browning of harvested longan fruit in association with membrane lipids metabolism [J]. Food Chemistry, 244: 93-101.

ZHANG Y, JIN P, HUANG Y, et al., 2016. Effect of hot water combined with glycine betaine alleviates chilling injury in cold-stored loquat fruit [J]. Postharvest Biol Technol, 118: 141-147.

ZHANG Z K, HUBER D J, RAO J P, 2011. Ripening delay of mid-climacteric avocado fruit in response to elevated doses of 1-methylcyclopropene and hypoxia-mediated reduction in internal ethylene concentration [J]. Postharvest Biology and Technology, 60: 83-91.

ZHANG Z, LI J, FAN L, 2019. Evaluation of the composition of Chinese bayberry wine and its effects on the color changes during storage [J]. Food Chemistry, 276: 451-457.

ZHANG Z, ZHANG H, QUAN R, et al., 2009. Transcriptional regulation of the ethylene response factor LeERF2 in the expression of ethylene biosynthesis genes controls ethylene production in tomato and tobacco [J]. Plant Physiology, 150 (1): 365-377.

ZHAO H, ZHANG S, MA D, et al, 2024. Review of fruits flavor deterioration in postharvest storage: Odorants, formation mechanism and quality control [J]. Food Research International, 182: 114077.

ZHAO S, JONES J A, LACHANCE D M, et al., 2015. Improvement of catechin production in Escherichia coli through combinatorial metabolic engineering [J]. Metabolic Engineering, 28: 43-53.

ZHAO Y T, ZHU X, HOU Y Y, et al., 2020. Postharvest nitric oxide treatment delays the senescence of winter jujube (zizyphus jujuba mill. cv. dongzao) fruit during cold storage by regulating reactive oxygen species metabolism [J]. Scientia Horticulturae, 261: 1-9.

ZHAO Y Y, SONG C C, BRUMMELL D A, et al., 2021. Jasmonic acid treatment alleviates chilling injury in peach fruit by promoting sugar and ethylene metabolism [J]. Food Chemistry, 338: 128005.

ZHAO Y, ZHU X, HOU Y, et al., 2021. Effects of harvest maturity stage on postharvest quality of winter jujube (Zizyphus jujuba Mill. cv. Dongzao) fruit during cold storage [J]. Scientia Horticulturae, 277.

ZHENG X, HU B, SONG L, et al., 2017. Changes in quality and defense resistance of kiwifruit in response to nitric oxide treatment during storage at room temperature [J]. Scientia Horticulturae, 222: 187-192.

ZHENG Z D, WANG T, LIU M Y, 2023. Effects of exogenous application of glycine betaine treatment on 'huangguoggan' fruit during postharvest storage [J]. International Journal of Molecular Sciences, 24 (18): 14316.

ZHONG Q P, XIA W S, 2007. Effect of 1-methylcyclopropene and/or chitosan coating treatments on storage life and quality maintenance of Indian jujube fruit [J]. LWT-Food Science and Technology, 40 (3): 404-411.

ZHOU D, LI L, WU Y, et al., 2015. Salicylic acid inhibits enzymatic browning of fresh-cut chinese chestnut (*Castanea mollissima*) by competitively inhibiting polyphenol oxidase [J]. Food Chemistry, 171: 19-25.

ZHOU X, DONG L, LI R, et al., 2015a. Low temperature conditioning prevents loss of aroma-related esters from 'Nanguo' pears during ripening at room temperature [J]. Postharvest Biology and Technology, 100: 23-32.

ZHOU Y, MA J, XIE J, et al., 2018. Transcriptomic and biochemical analysis of highlighted induction of phenylpropanoid pathway metabolism of citrus fruit in response to salicylic acid, Pichia membranaefaciens and oligochitosan [J]. Postharvest Biology and Technology, 142: 81-92.

ZHU F, CHEN J, XIAO X, et al., 2016. Salicylic acid treatment reduces the rot of postharvest citrus fruit by inducing the accumulation of H_2O_2, primary metabolites and lipophilic polymethoxylated flavones [J]. Food Chemistry, 207: 68-74.

ZHU L Q, DU H Y, WANG W, et al., 2019. Synergistic effect of nitric oxide with hydrogen sulfide on inhibition of ripening and softening of peach fruits during storage [J]. Scientia horticulturae, 256: 1-6.

ZHU Q G, WANG M M, GONG Z Y, et al., 2016. Involvement of DkTGA1 Transcription Factor in Anaerobic Response Leading to Persimmon Fruit Postharvest De-Astringency [J]. PLoS One, 11 (5): e0155916.

ZHU Z, LIU, R L, LI B Q, et al., 2013. Characterisation of genes encoding key enzymes involved in sugar metabolism of apple fruit in controlled atmosphere storage [J]. Food Chemistry, 141 (4): 3323-3328.

ZOZIO S, SERVENT A, CAZAL G, et al., 2014. Changes in antioxidant activity during the ripening of jujube (Ziziphus mauritiana Lamk) [J]. Food Chemistry, 150: 448-456.